Losing Strands in the Web of Life: Vertebrate Declines and the Conservation of Biological Diversity

JOHN TUXILL

Jane A. Peterson, *Editor*

WORLDWATCH PAPER 141

May 1998

THE WORLDWATCH INSTITUTE is an independent, nonprofit environmental research organization in Washington, DC. Its mission is to foster a sustainable society in which human needs are met in ways that do not threaten the health of the natural environment or future generations. To this end, the Institute conducts interdisciplinary research on emerging global issues, the results of which are published and disseminated to decisionmakers and the media.

FINANCIAL SUPPORT for the Institute is provided by the Geraldine R. Dodge Foundation, the Ford Foundation, the Foundation for Ecology and Development, the William and Flora Hewlett Foundation, W. Alton Jones Foundation, John D. and Catherine T. MacArthur Foundation, Charles Stewart Mott Foundation, the Curtis and Edith Munson Foundation, David and Lucille Packard Foundation, Rasmussen Foundation, Rockefeller Brothers Fund, Rockefeller Financial Services, Summit Foundation, Surdna Foundation, Turner Foundation, U.N. Population Fund, Wallace Genetic Foundation, Wallace Global Fund, Weeden Foundation, and the Winslow Foundation. The Institute also receives financial support from the Friends of Worldwatch and from our Council of Sponsor members: Toshishige Kurosawa, Kazuhiko Nishi, Roger and Vicki Sant, Robert Wallace, and Eckart Wintzen.

THE WORLDWATCH PAPERS provide in-depth, quantitative and qualitative analysis of the major issues affecting prospects for a sustainable society. The Papers are written by members of the Worldwatch Institute research staff and reviewed by experts in the field. Published in five languages, they have been used as concise and authoritative references by governments, nongovernmental organizations, and educational institutions worldwide. For a partial list of available Papers, see back pages.

Table of Contents

ACKNOWLEDGMENTS: For help with the preparation and review of this paper, I thank: Mark Bryer, David Olson, Sandra Postel, John Terborgh, and Julie Velasquez Runk for comments on various drafts; my Worldwatch colleagues Janet Abramovitz, Christopher Flavin, Hilary French, and Ashley Mattoon for their detailed reviews and help with background information; and Amie Brautigam at IUCN as well as Lori Brown, Brian Halweil, and Payal Sampat at Worldwatch for providing crucial statistics and references. I also am indebted to Liz Doherty for ably provided layout and design, as well as to Dick Bell, Mary Caron, Suzanne Clift, and Amy Warehime, for smoothly and skillfully managed production and outreach work. Special thanks to Jane Peterson for her careful editing—undiminished by the distances and deadlines involved; and to my Worldwatch colleague Chris Bright for his many contributions at all stages of this work.

JOHN TUXILL is a Research Fellow with the Worldwatch Institute, where he researches and writes about the conservation of biological diversity. He is a co-author of the Institute's 1998 reports *State of the World* and *Vital Signs*, and has also contributed to *World Watch* magazine. Currently he writes for Worldwatch from eastern Panama, where he is also involved in field research on the management of non-timber forest products. Prior to joining Worldwatch, he earned a master's degree in Conservation Biology and Sustainable Development from the University of Wisconsin, and worked on tropical forest management in Latin America and the Eastern Caribbean. He is also the author, with Gary Paul Nabhan, of a forthcoming book, *Plant Resources and Protected Areas: A Guide to In Situ Management,* to be published this year by Chapman & Hall.

Introduction

During the 1980s, biologists began to notice that frog populations around the world were declining drastically, even in regions that were well protected and seemingly pristine. Sharing study data at scientific meetings, they were perplexed and alarmed to find that frogs were disappearing simultaneously in many different settings. The scientists formed the Declining Amphibian Populations Task Force to investigate the declines in depth at field sites around the globe.

The evidence so far suggests these amphibians are under assault from a variety of hazards. Some species, like the mountain yellow-legged frog of California's Sierra Nevada, have vanished from their former haunts because of predation and competition from introduced species, such as non-native trout. Other frogs are suffering declines and developmental abnormalities that may be linked to exposure to fungicides and pesticides during their vulnerable egg and tadpole life stages. Excessive ultraviolet light levels from the Earth's thinning ozone layer appear to be damaging the eggs of frogs and toads that breed in high-altitude mountain ponds in the western United States. Frogs in the shady montane rainforests of Australia and Central America, on the other hand, are well protected from the sun's rays, but many of them are succumbing to virulent fungal infections.[1]

The fate of frogs is one of the clearest signs we have to date of a global decline in the environmental health of our planet—but they are not the only animals who can tell us about this unraveling. In Australia, 77 mammal species found nowhere else are currently threatened or already extinct, reflecting major changes in the terrestrial habitats of that continent. East Africa's Lake Victoria—the world's second-largest freshwater lake—appears to have lost over 60 percent of its 300-plus species of unique cichlid fishes beginning in the 1980s, following the intentional introduction of the predatory Nile perch. And scientists now believe that more than 2,000 bird species vanished in a wave of extinction as human populations first settled the Pacific island chains between 3,000 and 1,000 years ago—a progressive decline that continues even today on many archipelagos. The island of Guam, for instance, has lost 9 of its 12 native bird species over the past 20 years from predation by the non-native brown tree snake, and 2 of the remaining 3 are nearly gone.[2]

The ongoing biological impoverishment of our world often is most clearly seen in those organisms closest to us— birds, mammals, reptiles, amphibians, and fish. These are the vertebrate animals, distinguished from invertebrates by an internal skeleton and a spinal column—a type of anatomy that permits, among other things, complex neural development and high metabolic rates.

Although vertebrates are a relatively small taxonomic group—about 50,000 species—tracking their status is a huge task, and until recently we have known little about their overall health. Then, in the early 1990s, the World Conservation Union (also known as the IUCN) and collaborating organizations undertook an extensive review of the conservation status of animal species worldwide. Working through an international network of Species Specialist Groups, the IUCN consulted over 600 biologists with in-depth field knowledge of the animals concerned. The scientists, who represent many different countries and nationalities, assessed each species by a standard set of biological criteria that incorporated factors such as population size and trends, and changes in geographic distribution. (See Table 1.) The results of this global survey were published as the *1996 Red List*, a listing of all animal species known to be threatened with extinction around the world. The IUCN had published *Red Lists* before, but the 1996 version was unprecedented in its completeness and scientific rigor.[3]

The IUCN's sobering analysis suggests that one in every four vertebrates species is declining markedly, is limited to small populations, or is already threatened with extinction. The best-known vertebrate groups are birds and mammals, of whose species about 10 percent and 25 percent respectively are threatened with extinction. The IUCN effort has reviewed all bird and mammal species for their conservation status,

---

TABLE 1

Selected Criteria Used by IUCN to Determine the Conservation Status of Vertebrates

Criteria	Means of Evaluation
Populations of a species—no matter how large—are declining strongly.	• A past reduction in numbers is observed or inferred. • Present trends indicate falling population numbers that are likely to continue into the future.
The range of a species is small and shrinking.	• A species has disappeared from many areas where it used to occur. • A species is being fragmented into isolated sub-populations. • The extent or quality of a species' habitat is in obvious decline.
A species occurs in small populations that are declining (even if the decline is not strong).	• A species has a limited number of mature individuals in its populations. • A species is in a long-term or rapid decline, and only occurs in isolated sub-populations.
A species' total population is very small or very restricted in distribution (it does not necessarily have to be declining).	• A species has very few mature individuals. • A species only occurs in a very small geographic area.

Source: See endnote 3.

but the other vertebrate groups are not as well known—only about one fifth of reptile, one eighth of amphibian, and one tenth of fish species have been surveyed. According to these partial surveys, about 20 percent of reptiles and amphibians, and over one third of fishes, rank as threatened. The largest numbers of threatened vertebrates live in tropical regions, where species diversity is naturally highest, but even remote polar ecosystems harbor vulnerable species, such as the ivory-colored beluga whale and the spectacled eider, a little known duck from the Bering Sea. Generally speaking, larger-bodied mammals, birds, reptiles, and fish are declining much more sharply than smaller-bodied ones, a function of both their need for larger areas of undisturbed habitat and their appeal to hunters and fishermen.[4]

Vertebrate trends signal more than just the loss of individual species, of course. They are the most visible indicators of a decline in all facets of *biological diversity*, or "biodiversity" for short—a deceptively concise phrase that refers to the richness and complexity of life on Earth. Although biodiversity is most commonly measured as numbers of species, its true scope is much greater. It includes the distinct populations that make up a species in discreet areas, each with its own unique gene pool—which explains, for instance, why Siberian tigers in Russia are larger and more thickly furred than those in tropical India, even though they are the same species. Biodiversity also encompasses the ecological communities that form when species share a common habitat—a sedge meadow, an old-growth oak grove, a desert cactus forest. On an even larger scale, biodiversity is expressed as the landscape patterns created when communities assemble into entire ecosystems, such as the woodland and savanna patchwork of Tanzania's Serengeti Plains. Biodiversity also includes interactions and relationships between organisms—the many strategies that tropical trees have for securing pollinators or dispersing their seeds, or the intimate prey-predator cycles of lemming and snowy owl populations in the high Arctic. It even encompasses the vast array of physiological pathways and chemical compounds that organisms have evolved to obtain food, defend themselves from predators, and attract mates—the toxins of a poison-dart frog, the pheromones of a moth.[5]

This diversity of life evident today in all corners of the Earth is the result of over three billion years of evolution. From the beginning, population declines and species extinctions have been a natural part of this process, but there is something disturbingly different about current and recent extinction patterns. Examinations of the fossil record of marine invertebrates suggest that the natural or "background" rate of extinction—the rate that has prevailed over millions of years of evolutionary time—claims something on the order of one to 10 species per year. Scientists who study the fossil and archeological record of the recent geologic past—called the Quaternary Period—have accumulated substantial evidence to suggest that extinction rates have increased over the past several millennia. Most estimates of the current situation are that at least 1,000 species are lost per year, an extinction rate 100 to 1,000 times above the background rate even when calculated with conservative assumptions. Like the dinosaurs 65 million years ago, human society now finds itself in the midst of a mass extinction: a global evolutionary convulsion with few parallels in the entire history of life. Unlike the dinosaurs, however, we are not simply the contemporaries of a mass extinction—we are the reason for it.[6]

The loss of species and populations and the resulting simplification of the natural world touches everyone, for no matter where or how we live, biodiversity underpins our existence. The Earth's endowment of natural communities is the biological infrastructure that provides humanity with food, fibers, and many other products and "natural services" for which there are no substitutes. Biodiversity supports our health care systems; some 25 percent of drugs prescribed in the United States include chemical compounds derived from wild species, and worldwide the over-the-counter value of such drugs is at least $40 billion annually. Billions of people also rely on plant- and animal-based traditional medicine for their primary health care. Genetic diversity maintains the vigor of our crops and livestock—the rice harvested by millions of people around the world owes its disease resistance to genes transferred by plant breeders from a single wild rice species, *Oryza nivara*. Insects, birds, bats, and even lizards provide pollination services, without which we could not feed ourselves. Frogs, fish, and birds furnish natural pest control; mussels and other aquatic organisms cleanse our water supplies; plants and micro-organisms renew and enrich our soils. Healthy ecological landscapes filter and regulate our freshwater supplies, prevent soil erosion and flooding, and break down our sewage and industrial waste. Together, the global economic and environmental benefits of biodiversity are conservatively estimated to be at least $3 trillion per year, or equivalent to at least 11 percent of the annual world economic output.[7]

As we begin to appreciate the vast array of goods and services that natural systems provide, we are also beginning to realize that most of what we are losing is still a mystery. We have barely begun to decipher the intricate ecological relationships that keep natural communities running smoothly. We still do not know—even to a rough order of magnitude—how many species there are on Earth. To date, scientists have catalogued about 1.8 million species of animals, plants, fungi, bacteria, and other organisms; most estimates of the number yet to be formally described range from 4 to 40 million. (The single most species-rich group of organisms appears to be insects; beetles, in particular, currently account for 25 percent of all described species).[8]

This massive gap in our understanding of biodiversity makes it difficult to grasp the true dimensions of the current mass extinction—and mobilize effective responses to it. Our knowledge of patterns and levels of biodiversity losses remains relatively general—we can estimate the scale of species' disappearance, but more precise numbers on biodiversity losses are harder to come by. Likewise, without a solid grasp of most species' ecological relationships, we can only begin to assess what their disappearance might mean for our planet's life support systems—and for our own well-being.

Vertebrates help us over the hurdle of ecological ignorance in several ways. They provide a good-sized sample of organisms that inhabit nearly all environments on Earth, from the frozen expanses of Antarctica to scorching deserts and deep ocean abysses. By virtue of the attention they receive from researchers, vertebrates can serve as ecological bellwethers for the multitude of invertebrates, plants, fungi, and microbes that remain undescribed and unknown. Since vertebrates tend to have relatively large resource requirements and to occupy the top rungs in food chains, habitats healthy enough to maintain a full complement of native vertebrates will have a good chance of retaining the more obscure or cryptic organisms found there. Conversely, ecological degradation can often be read most clearly in native vertebrate population trends.[9]

Perhaps the most celebrated example of this "bellwether effect" was the intense research effort set off by the publication of Rachel Carson's *Silent Spring* in 1962, which described the danger that organochlorine pesticides like DDT pose to wild vertebrates, particularly birds. The study of wildlife toxicology that Carson pioneered is now routine: ecologists often monitor vertebrate populations as a way of checking on the general health of a natural community. In the North American Great Lakes, for example, researchers gauge water quality partly by examining the health of the fish. Some vertebrate declines, such as those of frogs, may signal trouble that we cannot clearly see in any other way.[10]

The magnitude of current vertebrate declines in the *Red List* confirms the gravity of our self-made biodiversity crisis; if we do not reverse the trends, we will be left with simplified natural communities and ecosystems less capable of providing the natural services we now take for granted. Human activities fueling vertebrate declines and attendant biodiversity losses include conversion, fragmentation, and disruption of native habitats (by far the most important factor); overexploitation of species for their meat, hides, horns, or medicinal or entertainment value; and aiding (intentionally and unintentionally) the spread of *invasive* species—highly adapt-able animals and plants that "hitch-hike" with humankind around the globe. Environmental contamination by synthet-ic chemicals and toxic pollution is a lesser problem, but has the potential to grow much larger in the near future. Ecological upheavals could also be wrought by global climate change within many of our lifetimes.

To reverse biodiversity losses, nations must strengthen existing institutions that offer solutions, such as national endangered species programs, and on an international level, the Convention on Biological Diversity. Some venerable approaches to conservation, such as national parks and nature reserves, need to be rethought and redesigned in cer-tain situations to make them work with, rather than against, local people who have long resided in a particular region or place. Governments need to follow through on their com-mitments to fund conservation efforts and to implement sound public policies for biodiversity conservation. And the international community needs to find better ways to relieve the debt burdens that continue to squeeze poor countries financially, constricting their options for sound natural resource management or social development.

Such advances will only be secure if we also commit ourselves to building a sustainable society—both in devel-oping countries, which contain the greatest concentrations of biodiversity, and in developed countries, which hold the most resources necessary for accomplishing positive change. This means establishing incentives for all countries to pur-sue ecologically sound and socially equitable management of their forests, waterways, fisheries, and agricultural endow-ments. It means supporting and expanding governments' efforts to reduce population growth, and transforming the consumption patterns that historically have resulted from economic growth and prosperity. And it means shaping our rapidly globalizing economy so that trade links support rather than degrade the natural systems that supply us with biodiversity's benefits.

Perhaps most challenging of all, we must accomplish

these changes while having only a sketchy understanding of the diversity of life and how nature works. We are entering the information age with a humbling lack of knowledge about the status of ecological systems—as eminent biologist Edward O. Wilson observes, we live on a barely explored planet. IUCN's pioneering effort to assemble the *Red List* is one example of the kind of information that is needed to take effective action on biodiversity loss and shape an ecologically literate citizenry. It is noteworthy that this database has been compiled by a non-governmental organization, with the direct financial support of only three governments—Great Britain, Taiwan, and the Sultanate of Oman. If the *Red List's* findings were discussed and debated as widely as annual gross domestic product figures or international capital flows, public pressure would leave government leaders no choice but to devote more effort to sustaining natural systems. Such a scenario could yet come to pass—but the first step is to learn what vertebrates have to tell us.[11]

The Winged Messengers

With their prominent voices, eye-catching colors and unparalleled mobility, birds receive a great deal of attention from both scientists and laypeople alike. As a result, we know more about the ecology, distribution, and abundance of the nearly 10,000 species of birds than any other class of organisms on Earth. Not surprisingly, birds were the first animals comprehensively surveyed for their conservation status, by the organization BirdLife International in 1992, followed by full reassessments in 1994 and 1996.[12]

The latest news is not good. An estimated two out of every three bird species are in decline worldwide, and about 11 percent of all birds are already officially threatened with extinction. (See Table 2.) Four percent—some 403 species—are "endangered" or "critically endangered," meaning they

TABLE 2

Conservation Status of Birds, 1996

Status	Total	Share
	(number)	(percent)
Total Number of Species Surveyed	9,615	—
Threatened		
Nearing Threatened Status	875	9
Vulnerable to Extinction	704	7
In Immediate Danger of Extinction	403	4

Source: See endnote 13.

are right on the brink of existence and could disappear through nothing more than a spell of bad luck, such as an unusually wet rainy season that coincides with a deadly outbreak of avian malaria or some other disease. These include species like the crested ibis, a graceful white wading bird that has been eliminated from its former range in Japan, Korea, and Russia, and is now down to one small population of only 22 individuals in the remote Qinling Mountains of China's Shaanxi Province. Another 7 percent of all birds are in slightly better condition in terms of numbers, annual breeding success, or range size, but still remain highly vulnerable to extinction.[13]

The red-cockaded woodpecker is one vulnerable species hopefully on the road to recovery. This small black-and-white striped bird is found only in mature pine forests of the southeastern United States—especially longleaf pine, a habitat levelled by logging and agricultural clearing over the past two centuries. The managed industrial forests that have sprung back up in the region during recent decades are too young and dense for the woodpeckers, who need large adult pines amid an open, park-like understory. The birds currently number about 5,000 pairs. Their continued recovery depends on the success of efforts to restore optimal longleaf pine habitat throughout the southeast United States by prescribed burning, replicating the once common low-intensity forest fires to which the pines and the woodpeckers are both adapted. Red-cockaded woodpeckers were originally placed on the U.S. Endangered Species List in 1973, but it is only in recent years that they have begun to show signs of increasing their populations.[14]

Membership in this pool of threatened species is not spread evenly amongst the different taxonomic orders or groups of birds. The most threatened major groups include rails and cranes (both specialized wading birds), parrots, largely terrestrial game birds (pheasants, partridges, grouse, and currassows), and pelagic seabirds (albatrosses, petrels, and shearwaters). About one quarter of the species in each of these groups is currently threatened. Only 9 percent of songbirds fall into this category, but they still contribute the single largest group of threatened species, 542, because they are far and away the most species-rich bird order.[15]

The leading culprits in the decline of birds are a familiar set of interrelated problems all linked to human activity: habitat alteration, overhunting and overcollecting, exotic species invasions, and chemical pollution of the environment. Habitat loss is by far the leading factor—over half of all threatened bird species are in trouble because of the transformation and fragmentation of old-growth forests, freshwater marshes, chaparral, semi-arid thornscrub, and other irreplaceable habitats. Of course, people have always modified natural landscapes in the course of finding food, obtaining shelter, and meeting other requirements of daily life. What makes present-day human alteration of habitat the number-one problem for birds and other creatures is its unprecedented scale and intensity.

Habitat loss can occur in a myriad of different patterns. Some are dramatic and obvious—China's Three Gorges Dam, if completed, will flood over 550 linear kilometers of one of the most biologically rich stretches of the Yangtze River; Brazilian logging companies have bulldozed a forest road over 500 kilometers long across the southern Amazonian state of Pará in search of mahogany trees. Other patterns unfold more subtly and gradually, as when marshes, seasonal wet meadows, "prairie potholes," and other small-sized wetlands across the United States are drained and plowed under to make room for expanded crops or suburban home tracts. In either case, when human use of a natural landscape expands or intensifies, natural habitats tend to shrink in size and become more fragmented and isolated—making it harder for the species that inhabit them to survive.[16]

The birds hit hardest by habitat loss are ecological specialists with small ranges. Such species tend to reside full-time in specific, often very local habitat types, and are most abundant in the tropical and subtropical regions of Latin America, Sub-Saharan Africa, and Asia. Over 70 percent of South America's rare and threatened birds do not inhabit lowland evergreen rainforests or the commonly cited hotspot of environmental concern, the Amazon Basin. Instead they hail from obscure but gravely disturbed habitats such as the cloud-shrouded montane forests and high-altitude wetlands of the northern and central Andes, deciduous and semi-arid Pacific woodlands from western Colombia to northern Chile, and the fast-disappearing native grasslands and riverine forests of southern and eastern Brazil. The long list of imperiled birds native to these little-noticed habitats signals that what is being lost in South America is not just rainforests, but a far more intricate ecological mosaic that is vanishing before most people have even become aware of its existence.[17]

High concentrations of gravely endangered birds are also found on oceanic islands worldwide. Birds endemic to insular habitats—that is, found nowhere else—account for almost one third of all threatened species, and an astounding 84 percent of all historically known bird extinctions. These numbers reflect the fact that island birds tend naturally to have smaller ranges and numbers, making them more susceptible to habitat disturbance. And since island birds are often concentrated in just a handful of populations, if one population is wiped out by a temporary catastrophe such as a drought, the birds often have few sources from which they can recolonize the formerly occupied habitat. Equally important is that many island birds have evolved in isolation for thousands or even millions of years, often becoming flightless and naively tame in the absence of predators. Such species are particularly vulnerable to human hunting, as well as predation and competition from non-native, invasive species. Highly adaptable animals and plants that spread outside their native ecological ranges—usually with intentional or inadvertent human help—invasives thrive in human-disturbed habitats.[18]

Scientists now know that island birds have had elevated extinction rates for several millennia. Archeological studies on central and south Pacific archipelagos indicate that these islands once were home to the world's most spectacular avian evolutionary showcase: flightless ibises and herons, oversized flightless geese, a host of honeycreepers and rails, and even—in New Zealand—giant emu-like birds called moas. These species first began to decline as human populations from what is now Southeast Asia and New Guinea island-hopped eastward across the Pacific between 3,000 and 1,000 years ago, settling the last large unoccupied habitable region on the planet. The ancestors of current-day Pacific Island peoples cleared and burnt habitats for agriculture, hunted native birds, and introduced non-native pigs, rats, and dogs. Since nearly every island chain held its own suite of unique endemic birds, scientists estimate that over 2,000 species—fully one sixth of the world's birds—disappeared from the Pacific. Subsequent European exploration and settlement of the region increased the magnitude of ecological disturbance on Pacific islands, which has advanced again in our modern age of global economic trade. As a result, island birds continue to dwindle.[19]

Pacific archipelagos were home to the world's most spectacular avian evolutionary showcase.

Among countries with more than 200 native bird species, the highest threatened percentage—15 percent—

occurs in two island archipelagos, New Zealand and the
Philippines. The tiny island nation of Mauritius in the
Indian Ocean has recorded 21 bird extinctions since the
arrival of humans in the 1600s. Mauritian species gone for-
ever include several species of flightless herons and the
famed dodo, an aberrant flightless pigeon nearly the size of
a turkey.[20]

No island birds, however, have been more decimated
than those of Hawaii. Nearly all of Hawaii's original 90-odd
bird species were found nowhere else in the world. Barely
one third of these species remain alive today, the rest having
vanished under Polynesian and modern-day impacts, and
two thirds of the remaining birds continue to be threatened
with extinction. The degree of ecological disruption in
Hawaii has been so great that virtually all lowland Hawaiian
songbirds are now non-native species introduced by
humans.[21]

While island species and habitat specialists dominate
the ranks of the world's most endangered birds, an equally
disturbing trend is population declines in more widespread
species, particularly those that migrate seasonally between
breeding and wintering grounds. In the Americas, more
than two thirds of the migratory bird species that breed in
North America but winter in Latin America and the
Caribbean declined in abundance between 1980 and 1991.
Some dropped by more than 4 percent per year, including
yellow-billed cuckoos, Tennessee warblers, and Cassin's
kingbirds. Two decades of bird surveys in Great Britain and
central Europe have also revealed serious declines in long-
distance migrants that winter in Sub-Saharan Africa.[22]

Long-term population declines in migratory birds are
tied to a host of contributing hazards. Habitat loss squeezes
species on both breeding and wintering grounds, as well as
at key stopping points—such as rich tidal estuaries for shore-
birds—along their migratory routes. The leading threat to
Europe's migratory—and non-migratory—bird species is the
abandonment of low-intensity agricultural practices, which
have long maintained hay meadows, upland pastures, rain-

fed grainlands, and other agricultural habitats that many native birds require for breeding. About 40 percent of European birds judged by BirdLife International to have "unfavorable" conservation status (indicating that their European populations are small and regional, or are substantially declining, or are highly localized) are dependent on traditional agricultural habitats. Since the 1960s, these habitats have been displaced by increasingly intensive agricultural practices, which have been strongly promoted by the high-yields-at-any-cost emphasis of Europe's Common Agricultural Policy (CAP).[23]

A growing European public interest in organic agriculture, however, could help restore and revive agricultural habitats for birds' benefit. One trial in Britain compared wintering birds on an unplowed field (seeded with a no-till "direct-drilling" method) with those on a standard autumn-plowed field, both of which had similar grain yields. Observers tallied only seven skylarks (a vulnerable bird species in Europe) in the plowed field, but the unplowed field held 157 skylarks—plus an additional 117 tree sparrows, 35 chaffinches, and 159 yellowhammers, a clear sign that the unplowed field mimicked traditional farmland habitat. The CAP is currently under review by the European Commission with an aim to revamping long-standing subsidies, but so far the commission has made no move to incorporate agricultural standards, organic or otherwise, that could bring habitat relief for Europe's dwindling farmland birds.[24]

Excessive hunting also remains a hazard for many migratory species. In many Mediterranean nations, there is an enduring tradition of pursuing all birds indiscriminately, regardless of size or status. Local species are hunted intensively for food, and migrants that breed in northern Europe must brave an annual fusillade of guns and snares as they fly south to Africa. In Italy alone, as many as 50 million songbirds are harvested every year as bite-sized delicacies.[25]

Exposure to chemical pollution is another problem that birds face. Birds are at greatest risk from pesticide and pollution exposure in developing countries, where many chemicals banned from use in industrial nations continue to be applied or discharged indiscriminately. In late 1995 and early 1996, about 5 percent of the world's population of Swainson's hawks—some 20,000 birds—died in unintentional mass poisonings on their wintering grounds in Argentina's pampas. Argentine farmers were applying heavy doses of an internationally manufactured organophosphate pesticide called monocrotophos to control grasshopper outbreaks on their crops. The hawks, which breed in western North America, were exposed to the chemical when they fed on the grasshoppers, one of their main winter food sources. Argentina has since banned the use of monocrotophos on grasshoppers and alfalfa, and no large hawk kills were found during the 1996-97 wintering season, but it is unclear how long it will take Swainson's hawk populations to recover from these substantial losses.[26]

Whether reduced by the conversion of key habitats such as wetlands, by overkill in the form of hunters' guns, or by toxic contamination of water and food supplies, the decline of migratory birds is most sobering because it is a loss not just of individual species, but of an entire ecological phenomenon. Present-day migrants must negotiate their way across thousands of kilometers of increasingly tattered ecological landscapes. The fact that many birds continue to make this journey, despite the obstacles, is cause for hope and inspiration. Yet as long as bird diversity and numbers continue to spiral downward, there can be no rest in the effort to protect breeding grounds, wintering areas, and key refueling sites that all birds—migratory and resident—simply cannot live without.

Twilight for Mammals

When the conservation status of birds was first comprehensively assessed, the degree of endangerment—about 11 percent—was taken as the best available estimate of endangerment for all vertebrates, invertebrates, and other life on Earth. Then, in 1996, the IUCN comprehensively reviewed the status of all mammal species for the first time, allowing for a full comparison with birds. The news was disconcerting—about 25 percent of all mammal species are treading a path that, if followed without intervention, is likely to end in their disappearance from Earth. (See Table 3.) This suggests that mammals are substantially more threatened than birds, and raises a larger question about which group better represents the level of endangerment faced by other, less well studied organisms.[27]

Out of almost 4,400 mammal species, about 11 percent are already endangered or critically endangered. Another 14 percent of mammals remain vulnerable to extinction, including the Siberian musk deer, whose populations in Russia, though still extensive, have fallen 70 percent during this decade due to increased hunting to feed the booming trade in musk, used in perfumes and traditional Asian medicine. An additional 14 percent of mammal species also come very close to qualifying as threatened under the criteria used by IUCN to assess species' status. These "near-threatened" species tend to have larger population sizes or be relatively widespread, but nonetheless face pressures that have set them on the fast track to threatened status in the

TABLE 3

Conservation Status of Mammals, 1996

Status	Total (number)	Share (percent)
Total Number of Species Surveyed	4,355	–
Nearing Threatened Status	598	14
Threatened		
Vulnerable to Extinction	612	14
Immediate Danger of Extinction	484	11

Source: See endnote 27.

not-too-distant future. One near-threatened species is the African red colobus monkey, which occurs in scattered populations all the way across the continent, from Senegal to Kenya, and encounters hunting pressure and habitat loss everywhere.[28]

Among major mammalian groups, primates (lemurs, monkeys, and apes) occupy the most unfortunate position, with nearly half of all primate species threatened with extinction. Also under severe pressure are hoofed mammals (deer, antelope, horses, rhinos, camels, and pigs) with 37 percent threatened, insectivores (shrews, hedgehogs, and moles) with 36 percent, and marsupials (opossums, wallabies, and wombats) and cetaceans (whales and porpoises) at 33 percent each. In slightly better shape are bats and carnivores (dogs, cats, weasels, bears, raccoons, and hyenas) at 26 percent apiece. At 17 percent, rodents are the least-threatened mammalian group, but also the most diverse. Like songbirds, rodents still contribute the most threatened species—300—of any group.[29]

The biggest source of loss of mammalian diversity in the late twentieth century is the same as that for birds—habitat loss and degradation. As humankind converts old-growth forests, grasslands, riverways, and wetlands, for intensive agriculture, tree plantations, industrial development, and transportation networks, we relegate many mammals to precarious existences in fragmented, remnant habitat patches that are but ecological shadows of their former selves.

Habitat loss is a principal factor in the decline of at least three quarters of all mammal species, and it is the only significant factor for many small, cryptic rodents and insectivores that are not directly persecuted. The main reason primates are so threatened is that they are dependent on large expanses of tropical forests, a habitat under siege around the globe. In regions where tropical forest degradation and conversion have been most intense, such as South and East Asia, Madagascar, and the Atlantic forest of eastern Brazil, on average 70 percent of the endemic primate species face extinction.[30]

The loss of habitat also afflicts marine mammals, though it usually proceeds as gradual, cumulative declines in habitat quality rather than wholesale conversion of an ecological landscape (as when a forest is cut down and replaced by a housing development). Marine mammals, particularly those that inhabit densely populated coastal areas, now have to contend with polluted water and food, physical hazards from fishing gear, heavy competition from humans for the fish stocks on which they feed, and hazardous, noisy boat traffic. Along the coastline of Western Europe, bottlenose dolphins and harbor porpoises—the only two cetaceans that regularly use nearshore European waters—seem to be steadily declining. Seal populations in the Baltic Sea carry very high chemical pollutant loads in their tissues that appear to decrease their reproductive success.[31]

In addition to habitat loss, at least one in five threatened mammals faces direct overexploitation—excessive hunting for meat, hides, tusks, and medicinal products, and persecution as predators of, and competitors with, fish and livestock. Larger mammals tend to be overhunted much more frequently than smaller ones, and when strong market demand exists for a mammal's meat, hide, horns, tusks, or bones, species can decline on a catastrophic scale.[32]

While the drastic population crashes of great whales, elephants, and rhinos are well known, the threat of overexploitation reaches much further. For instance, only the most remote or best-protected forests throughout Latin America have avoided significant loss of tapirs, white-lipped peccaries, jaguars, wooly and spider monkeys, and other large mammals that are heavily hunted. The result is what ecologists call the "empty forest effect"; the trees may still be standing, but the wildlife has vanished.

Most hunting in Latin America is done for home subsistence—wild game meat is an important source of protein in the diets of rural residents, particularly for indigenous people. One estimate pegs the annual mammal take in the Amazon Basin at over 14 million individuals. As large as the Amazon haul is, it is not an entirely unmanaged harvest.

Many indigenous societies which depend on wildlife for subsistence—both in the Amazon and the world over—have traditions and beliefs that help limit their depletion of this food source. Some cultures designate certain animals as taboo to hunt or eat—a practice that is commonly applied to certain primates, such as langurs in India and some species of lemurs in Madagascar.[33]

The Aché people of eastern Paraguay are one society that does not follow taboos—they depend heavily on wildlife for their food supply, and hunt all larger mammals, birds, and reptiles that occur in the region's forests. However, up until the 1970s, the Aché lived in extended family groups that were highly nomadic, rarely spending more than a few days in one spot. This dispersed their hunting pressure across a very wide area, reducing the odds that the Aché would extinguish any of their prey species at any given site. Unfortunately, traditional practices that protect wildlife from overexploitation are falling by the wayside in modern times. Younger generations no longer attach as much value to the old traditions, subsistence patterns shift, human population densities increase, and long-term residents increasingly must compete with outsiders and newcomers for a limited pool of game animals. The Aché, for their part, now live in settled communities on reservations that include only a fraction of their former territory. As a result, Aché hunters—and their Paraguayan colonist neighbors—have greatly depleted wildlife populations in the forest areas that remain.[34]

The most severe overexploitation tends to occur when hunting is done to supply markets rather than just to feed hunters' families. Studies of communities in Peru and Gabon suggest that when they can sell it to markets, people harvest over twice as much wildlife as they might take for subsistence alone. In central African forests, there is now intensive, indiscriminate hunting of many species for the regional trade in wild game or bushmeat. In parts of Cameroon, the Republic of Congo, and other countries, the sale of bushmeat to traders supplying urban areas is the main income-generating activity available to rural residents. Rural and urban bushmeat consumption in Gabon has been estimated at 3,600 tons annually. The bushmeat trade is closely linked in many areas with internationally financed logging operations, which open up roads into previously isolated areas, thereby giving hunters access to new game-rich territory. In some cases, timber company employees boost their own income by providing hunters with ammunition and transportation on company vehicles in exchange for a share of the proceeds earned by selling the meat.[35]

Throughout South and East Asia, a major factor driving excessive wildlife exploitation is the demand for animal parts in traditional medicine. Tigers—the largest of all cats—once ranged from Turkey to Bali and the Russian Far East, and have been the subject of organized conservation projects for over two decades. At first these projects appeared to be having some success—until the mid-1980s brought a burgeoning demand for tiger parts for aphrodisiacs and medicinal products in East Asia. With the bones, hide, and other body parts of a single tiger potentially worth as much as $5 million, illegal hunting skyrocketed, particularly in the tiger's stronghold—India. Wild tigers now total at most 3,000 to 5,000 individuals, many in small, isolated populations that will be hard pressed to survive the many pressures they face.[36]

Valuable timber trees failed to regenerate normally after the crash of elephant populations.

The extirpation of a region's top predators or dominant herbivores is particularly damaging because it can trigger a cascade of disruptions in the relationships among species that maintain an ecosystem's diversity and function. Large mammals tend to exert inordinate influence within their ecological communities by consuming and dispersing seeds, creating unique microhabitats, and regulating populations of prey species. In Côte d'Ivoire, Ghana, Liberia, and Uganda, certain trees—including valuable timber species—

have failed to regenerate normally after the crash of elephant populations, upon whom the trees depend for seed dispersal. Similarly, decades of excessive whaling until the 1980s reduced the number of whales that die natural deaths in the open oceans. This may have adversely affected unique deep-sea communities of worms and other invertebrates that decompose the remains of dead whales after they have sunk to the ocean floor.[37]

One striking experiment demonstrating the important ecological role of top or "keystone" predators is now unfolding in Yellowstone National Park, in the western United States. Wolves had been absent from Yellowstone for about 50 years before they returned in 1995—both through an official government reintroduction program and by recolonizing the park themselves from Canada. The wolves—now numbering nearly 100—have subsisted primarily on elk, a large deer that many biologists felt had become unnaturally abundant in Yellowstone and were overbrowsing much of the park's best wildlife habitat. In order to elude the wolves, elk herds are now spending more time on higher ground where they can spot wolves more easily. Ecologists expect this shift will promote the recovery of rich river bottom stands of willows and aspen that had been bearing the brunt of the elks' appetites.[38]

The Yellowstone wolves are also dramatically reducing the abundant coyote population of the park; total coyote numbers are expected to drop by about two thirds. Wolves generally do not tolerate the smaller-bodied coyotes in their territory, and will often chase them away or kill them. Coyotes primarily hunt small rodents, and prior to the wolves' arrival they accounted for about 75 percent of Yellowstone's voles, ground squirrels, and pocket gophers that were eaten by predators annually. With wolves reducing the coyotes, this rodent supply is now available to other predators, including eagles, hawks, owls, badgers, and pine martens—a change expected to promote a more balanced and thus more diverse ecological community. And not all coyotes are losing out; those that manage to persist on the edges of wolves' territories are flourishing, because they now have an extra food source available—leftover elk kills. Because wolves usually only eat a portion of the meat on an elk carcass, elk kills are a bonanza as well for other part-time scavengers such as eagles, ravens, and Yellowstone's most endangered inhabitant, the grizzly bear. Scientists suspect that this newly abundant food source may even help grizzlies boost their population. Bear cubs are born during the mother's hibernation, and the number of cubs a female grizzly produces is directly dependent on her nutritional condition when she enters her winter den.[39]

Mammals in most regions have been less susceptible than birds to invasive species, but there is one big exception—the unique marsupial and rodent fauna of Australia, long isolated from other continents. The introduction of non-native rabbits, foxes, cats, rats, and other animals has combined with changing land-use patterns during the past two centuries to give Australia the world's worst modern record of mammalian extinction. Nineteen mammal species have become extinct since European settlement in the 18th century, and at least one quarter of the remaining native mammalian fauna are still threatened. Most declines and extinctions have occurred among small to medium-sized ground-dwelling mammals, such as bandicoots, hare wallabies, and mice, from interior Australian drylands. These habitats have been drastically altered by invasive species (particularly rabbits) in conjunction with extensive livestock grazing, land clearance for wheat cultivation, and altered fire patterns following the decline of traditional aboriginal burning of brush and grasslands.[40]

Taken together, the problems bedeviling mammals in today's world—habitat loss, overhunting, invasive species—are essentially the same as those faced by birds. So how can we account for the fact that one in every four mammals is in danger of extinction, versus only one in every 10 birds? The answer may be found in how well mammals and birds cope with the pressures placed upon them by humankind. Since birds tend to be more mobile and wide ranging, they may be able to find food and shelter more easily in the fragmented and disjointed landscapes produced by human disturbance. Birds are also smaller on average than mammals, so they require smaller ranges and fewer resources for survival—advantages when habitat and food supply become restricted. But while few other organisms have the resource demands of most mammals, few likewise are as mobile as birds, making it difficult to predict which group is a better guide for assessing the level of endangerment of other organisms.

The Hidden Fauna

Like their furred and feathered vertebrate kin, reptiles and amphibians (known collectively to scientists as *herpetofauna*) do not possess huge numbers of species—about 6,300 are documented for reptiles and 4,000 for amphibians. Nonetheless, both groups share with the world's many invertebrates the fate of being less well known and relatively little studied. To date, only a fifth of all reptile species and barely one eighth of all amphibian species have been formally assessed by scientists for their conservation status. Among reptiles, the conservation status of turtles, crocodilians, and tuataras (an ancient lineage of two lizard-like species living on scattered islands off New Zealand) has been comprehensively surveyed. But most snakes and lizards remain unassessed, as do the two main orders of amphibians, frogs, and salamanders.[41]

The herpetofauna that have been surveyed reveal a level of endangerment closely in line with that of mammals. (See Tables 4 and 5.) Twenty percent of surveyed reptiles currently rank as endangered or vulnerable, while 25 percent of surveyed amphibians receive the same designation. The country with the highest number of documented threatened herpetofauna is Australia at 62 species, followed closely by the United States with 52 species. These are not the most species-rich countries for these creatures—Brazil leads in amphibians and Mexico has the most reptiles—they have simply been more thoroughly surveyed and monitored.[42]

Among reptiles, species are declining for reasons similar to those affecting birds and mammals. Habitat loss is again the leading factor, contributing to the decline of 68 percent of all threatened reptile species. In island regions, habitat degradation has combined with exotic species to fuel the decline of many endemic reptiles. In the Galápagos Archipelago, the largest native herbivores are reptiles—long-isolated giant tortoises and land and marine iguanas found nowhere else in the world. Introduced goats are winning out over the native reptiles, however, and these interlopers have already eliminated unique populations of tortoises on 3 of 14 islands within the Galápagos chain. At least two other tortoise populations are in imminent danger.[43]

In addition, 31 percent of threatened reptiles are affected directly by hunting and capture by humans. This figure may be somewhat lower for reptiles since the groups most thoroughly assessed to date—turtles and crocodilians—are also among those most pursued by humans. Nevertheless, the high number also clearly reflects the heavy exploitation suffered by reptiles at the hand of collectors who seek them for the pet trade.[44]

TABLE 4

Conservation Status of Reptiles Surveyed, 1996

Status	Total (number)	Share (percent)
Total Number of Species Surveyed[1]	1,277	—
Nearing Threatened Status	79	6
Threatened		
Vulnerable to Extinction	153	12
In Immediate Danger of Extinction	100	8

[1]This survey includes only about one fifth of the 6,300 reptile species that are estimated to exist.

Source: See endnote 42.

TABLE 5

Conservation Status of Amphibians Surveyed, 1996

Status	Total (number)	Share (percent)
Total Number of Species Surveyed[1]	497	–
Nearing Threatened Status	25	5
Threatened		
Vulnerable to Extinction	75	15
In Immediate Danger of Extinction	49	10

[1]This survey includes only about one eighth of the 4,000 amphibian species that are estimated to exist.

Source: See endnote 42.

The plight of sea turtles has been studied and publicized since at least the 1960s, and all seven species are judged by IUCN as endangered. Although there has been progress on protecting sea turtles at some of their best-known nesting grounds, excessive harvest of these reptiles for meat and eggs remains a widespread problem. Where beaches are lit at night with artificial lights, as at tourist resorts, hatchling turtles become disoriented and crawl towards the land rather than the sea. Moreover, sea turtles continue to suffer inadvertent but significant mortality from nets set for fish—which entangle and drown the air-breathing turtles—and from shrimp trawlers, which scoop up huge numbers of turtles and fish as a "by-catch." Virtually all turtles can escape shrimp trawls that are equipped with a trap door-like device called a "turtle-excluder," but most shrimpers have been loath to install them except when mandated to do so by governments. U.S. shrimp trawlers have been required to use the devices since 1992, saving an estimated 55,000 sea turtles annually. Some Latin American nations also require their use, but Asian governments have not followed suit, claiming that turtle excluders are too expensive for their shrimp fleets to adopt.[45]

Although less well known than their seagoing relatives,

tortoise and river turtle species also are exploited intensive-ly in certain regions, to the point where many populations are greatly depleted. Tortoises and river turtles throughout Southeast Asia have long been an important source of meat and eggs for local residents. There is now also a growing international trade in these species to China, where they are used in traditional medicine. According to a recent report by TRAFFIC, a group that monitors the international wildlife trade, the annual East Asian trade in tortoises and river tur-tles involves some 300,000 kilograms of live animals, with a value of at least $1 million. At least five turtle species includ-ed in this trade are now candidates for the most stringent listing available under the Convention on International Trade in Endangered Species (CITES), which regulates inter-national wildlife trade. The strictest listing would place the turtles on "Appendix 1" of CITES, which requires that gov-ernments must establish import and export quotas and have them scientifically verified as non-harmful, for any trade in a listed species to occur. Traders must obtain import and export permits for any species so listed, and also must demonstrate that they have not obtained the animals ille-gally in violation of domestic laws of the source country.[46]

Certain species of crocodilians are still overhunted (such as black caimans in the Amazon Basin) and suffer from pollution (such as the Indian gharial and the Chinese alliga-tor), but this is one of the few taxonomic groups of animals whose overall fate has actually improved over the past two decades. Since 1971, seven alligator and crocodile species have been taken off IUCN's *Red List*, including Africa's Nile crocodile and Australia's huge estuarine crocodile. In part, these recoveries are due to the development of crocodile ranching operations, which harvest the animals for their meat and hides; when combined with effective wildlife pro-tection efforts, ranching can take hunting pressure off wild populations. In Zimbabwe, crocodile ranches have been so successful that domestic crocodiles now outnumber the country's 50,000 wild crocs by three to one. In 1991, croco-dile farming worldwide generated over $1.7 million in international trade.[47]

For amphibians in general, direct exploitation is less of a problem. With the exception of large frogs favored for their tasty legs, few amphibians face any substantial hunting pressure. Habitat loss remains a serious problem, however, affecting some 58 percent of threatened amphibians, mainly through the drainage, conversion, and contamination of wetland habitats. In addition, the spread of road networks and vehicular traffic leads to increased amphibian mortality that can decimate local amphibian populations.[48]

Habitat loss alone, however, cannot account for the drastic amphibian declines—and even extinctions—that have captured worldwide attention since the early 1990s. These baffling reversals have been particularly well documented among frogs in little-disturbed mountain habitats in Central America and the western United States, as well as in 14 species of rainforest-dwelling frogs in eastern Australia. Scientists are making good progress on identifying the immediate factors causing these declines. They include introduced predators, particularly gamefish like bass and trout; increases in ultraviolet radiation, which inhibit egg development; unusual climatic fluctuations, such as extended drought; and the lingering effects of past and present chemical contamination. The cause of population crashes of tropical montane frogs—previously the most mysterious of all—has been recently confirmed as a virulent fungal infection that strikes at stream-breeding frog species. But many declines may be explained best by synergistic combinations of these factors. For instance, the 1989 population implosions of frogs at the Monteverde Cloud Forest preserve in Costa Rica—including the famed golden toad, now extinct—correlated closely with an unprecedented drought, and also followed the pattern of other Central American montane frog communities hit by the infectious fungus. This suggests that the stress placed on frogs by the drought conditions may have predisposed them to fungal attack.[49]

Yet even where researchers have identified the proximate problems spurring frog declines, the ultimate causes are still murky. For instance, the fungus attacking tropical montane frogs is called a chytrid fungus, and is an ubiquitous component of tropical streamside habitats; why it is suddenly able to attack frogs who have presumably lived alongside it for millennia is still a mystery. It may be that frogs, with their highly permeable skins and life cycles dependent on both aquatic and terrestrial habitats, are indicating—more clearly than any other group of organisms—the gradual decline of our planet's environmental health.[50]

Under the Waters

The world's fish offer the best measure of the state of biological diversity in aquatic communities. Fish occur in nearly all permanent water environments, from the perpetually dark ocean abyss to isolated alpine lakes and alkaline desert springs. Fish are also far and away the most diverse vertebrate group—nearly 24,000 fish species have been formally described by scientists, about equal to all other vertebrates combined.[51]

As with reptiles and amphibians, less than 10 percent of fish species have been formally assessed for their conservation status, with marine fish (some 14,000 species) being notably understudied. This partial assessment brings particularly disturbing news, however, for the numbers suggest that fully one third of all fish species are already threatened with extinction. (See Table 6.)[52]

The causes of fish endangerment—habitat alteration, exotic species, and direct exploitation—are no different from those afflicting other species, but they appear to be more pervasive in aquatic ecosystems. Freshwater hotspots of fish endangerment tend to be large rivers heavily disturbed by human activity (such as the Missouri, Columbia, and Yangtze rivers), and unique habitats that hold endemic fish faunas, such as tropical peat swamps, semi-arid stream systems, and isolated large lakes. Saltwater hotspots include

TABLE 6

Conservation Status of Fish Surveyed, 1996

Status	Total (number)	Share (percent)
Total Number of Species Surveyed[1]	2,158	—
Nearing Threatened Status	101	5
Threatened		
Vulnerable to Extinction	443	21
In Immediate Danger of Extinction	291	13

[1]This survey includes only about one tenth of the 24,000 fish species that are estimated to exist.

Source: See endnote 52.

coral reefs, estuaries, and other shallow, near-shore habitats.[53]

Although degradation of terrestrial habitats such as forests may be more obvious and get the most attention, freshwater aquatic habitats receive an even heavier blow from humanity. More than 40,000 large dams and hundreds of thousands of smaller barriers plug up the world's rivers, altering water temperatures, sediment loads, seasonal flow patterns, and other river characteristics to which native fish are adapted. Levees disconnect rivers from their floodplains, eliminating backwaters and wetlands that are important fish spawning grounds. The effects of river engineering works also surface in distant lakes and estuaries, whose ecologies decline when river inflows are altered. Agricultural and industrial pollution of waterways further reduces habitat for fish and other aquatic life. Agricultural runoff in the Mississippi River Basin is now so extensive that when the river enters the Gulf of Mexico, the over-fertilized brew of nutrients it carries sparks huge algal blooms, which deplete the water of oxygen and create a "dead zone" of some 17,600 kilometers—nearly the size of New Jersey.[54]

As a result of all these problems, at least 60 percent of threatened freshwater fish species are in decline because of habitat alteration. These include 26 species of darters—small, often brightly colored fish that used to be common in the now heavily dammed rivers of the southern United States—and 59 threatened species of fish in India recently identified in a nationwide study by the Zoological Survey of India.[55]

Alteration of aquatic habitats has been particularly catastrophic for native fish in semi-arid and arid regions, where human competition for water resources is high. In the heavily altered Colorado River system of the arid southwestern United States and northwest Mexico, 29 of 50 native fish species are either extinct or endangered. This includes the totoaba, a marine fish that used to breed in the Colorado River Delta—in most years now the river runs dry well before it reaches the sea. Elsewhere in semiarid areas of Mexico, river and spring systems have lost an average of 68 percent of their native and endemic fish species because of falling water tables and altered river hydrologies, both due to the water demands of a growing human population.[56]

The world's coral reefs are thought to be second only to tropical rainforests in species diversity, and are being damaged just as rapidly. Reefs cover 600,000 square kilometers of shallow coastal waters and by some estimates hold over 1 million species of animals and plants. The biggest threat to reefs is degradation by pollution and sedimentation from coastal development. Nutrient-rich agricultural runoff and sewage wastes deplete coastal waters of oxygen, thus encouraging the growth of algae on reefs, which eventually smothers the corals. Reefs are also physically damaged by mining of coral for construction projects, and also by fishermen who toss dynamite overboard (known as "blast fishing") and squirt cyanide into coral grottoes to catch their quarry. The latter technique is now rampant in the Indo-Pacific region (East Africa, South Asia, and Southeast Asia) because the cyanide only stuns the fish, allowing them to be captured alive. Live reef fish fetch a premium price in the aquarium trade and in Southeast Asian and Chinese restaurants as well. Together these markets are estimated to consume 20,000 to 25,000 tons of live fish annually.[57]

Once stressed by sedimentation, pollution, and ecological imbalances, coral reefs become more susceptible to damage from hurricanes and diseases, which can in turn degrade them further. All told, 30 percent of all reefs worldwide are thought to be in critical condition, with 10 percent of reef ecosystems already in a state of collapse. In a global reef survey called Reef Check, conducted in 1997 at 300 sites in the Caribbean, the Indo-Pacific, and the Red Sea, 95 percent of the reefs examined showed significant habitat degradation and loss of indicator species such as groupers and lobsters. In Jamaica, studies documented how healthy coral cover on reefs plunged from 52 percent in 1980 to only 3 percent in 1993. Indonesian researchers working near Jakarta must now motor 25 kilometers out from shore to find any live coral at all—and to encounter a healthy reef they must go even further. Only in well-protected marine parks (like Australia's 2,000-kilometer-long Great Barrier Reef) and in community-supported reef reserves (such as the Apo Island reef sanctuary in the Philippines) are reefs holding their own. Fortunately, countries such as Belize, Sri Lanka, and Thailand have begun to address the degradation of their reefs through meaningful government policy changes that aim to ban coral mining, control tourism in reef zones, and implement national reef conservation strategies.[58]

In an aquatic mirror of terrestrial islands, isolated freshwater lakes and river ecosystems are very susceptible to invasive species. Introductions of non-native, often predatory fish can unravel diverse native fish assemblages in just a few years, precipitating a cascade of local extinctions. Some 34 percent of threatened freshwater fish face pressure from introduced species, but none have been more devastated than the native cichlids of East Africa's Lake Victoria, the world's second-largest freshwater lake. The cichlid community was extraordinarily diverse, with over 300 specialized species, 99 percent of which occurred only in this lake. Unfortunately the community began to collapse during the 1980s following a population explosion of the Nile perch, a non-native predatory fish introduced to boost the lake fisheries. It did its job all too well, feeding indiscriminately on the much smaller cichlids and destroying native food webs. As many as 60 percent of the Lake Victoria cichlids may now be extinct, with only a museum specimen and a scientific name to mark their tenure on our planet.[59]

Many fish species also face a high degree of exploitation from commercial fisheries, particularly marine fish and species like salmon that migrate between salt and fresh water. About 68 percent of all threatened marine species suffer from overexploitation. The days when experts thought it impossible to deplete marine fish populations are long since gone, and scientists now realize that overexploitation is a serious extinction threat for many ecologically sensitive species.[60]

Thirty percent of all coral reefs worldwide are in critical condition.

Take seahorses, for example, which are captured for use in aquariums, as curios, and in traditional Chinese medicine. The global seahorse trade is very lucrative—top-quality dried seahorses have sold for up to $1,200 per kilogram in Hong Kong. Current worldwide seahorse harvests may top 20 million animals annually, and in China alone, demand is rising at almost 10 percent per year. Seahorses are unlikely to support such intensive harvesting for long because of their basic biology. Seahorses have low reproductive rates and complex social behavior (they are monogamous, with males rearing the young); they live in accessible habitat (shallow, inshore waters); and their low mobility makes them easy to catch. Already, some 36 seahorse species are threatened by the growing, unregulated harvest.[61]

Sharks are a second group of marine fish headed for trouble. As top ocean predators, sharks tend to be sparsely distributed, and most species grow and reproduce quite slowly. They are valued for their skin, meat, cartilage (reputed to have anti-cancer properties), liver oil, and especially fins, which are one of the most highly valued seafood commodities due to their popularity in East Asian cuisine. Reported worldwide shark catches have been increasing steadily since the 1940s, topping 730,000 tons by 1994. Unreported and incidental catches likely push that figure much higher, and most harvested shark species are probably already declining.[62]

Other fish have supported commercial fisheries for centuries but now appear unable to continue doing so in the face of additional threats from habitat alteration and pollution. Sturgeon, one of the most ancient fish lineages, occur in Europe, North and Central Asia, and North America. They have long been harvested for their eggs, famous as the world's premier caviar. Russia and Central Asia are home to 14 sturgeon species—the largest aggregation in the world—and produce 90 percent of the world's caviar, mostly from the Caspian Sea region. The sturgeon fishery was relatively well regulated during the Soviet era, but massive water projects and widespread water pollution led to sturgeon population crashes, so that all 14 sturgeon species are now endangered. To compound the problem, sturgeon poaching has been rampant the past several years due to minimal enforcement of fishing regulations in the post-Soviet Central Asian nations bordering the Caspian Sea—only Iran has had a controlled harvest. The total Caspian sturgeon population was estimated to have fallen 70 percent between 1978 and 1994, at which point it totaled about 43 million fish distributed among 6 sturgeon species. In 1995, over 90 percent of the sturgeon harvest was thought to be illegal.[63]

Continued uncontrolled exploitation of the stocks that remain would probably be the final nail in the coffin for these magnificent fish—but there are a few glimmers of hope. At the end of 1996, caviar industry leaders and government fishery committees from Russia, Kazakhstan, Turkmenistan, Azerbaijan, and Iran reached an agreement to ban all open-sea fishing for sturgeon on the Caspian Sea. The agreement specified that sturgeon fishing would only be legal on the lower Volga and Ural rivers, and that all five countries would undertake raids to apprehend poachers. These advances are welcome news, but sturgeon are slow-growing fish, and protecting the fishery will be a correspondingly long-term commitment for Central Asian nations. One study estimates it will take at least 25 to 30 years to rebuild Caspian Sea breeding stocks to levels stable enough to permit a sustainable harvest once again. And with sturgeon caviar selling at $80 per ounce in the United States and consumption apparently on the rise, incentives for poachers will likely remain strong without a commitment from caviar-consuming nations to help separate out the legal and illegal trade.[64]

With the collapse of native fish faunas in many river basins and lake systems, and growing awareness that certain groups of marine fishes are in decline, the evidence suggests that biological diversity is faring no better underwater than on land. One out of every three fish species surveyed now appears to be on the path to extinction. If this percentage holds up as more fish species are reviewed for their conservation status, it portends a grim future for other aquatic life on Earth.

Scaling Up to the Big Picture

Together, birds, mammals, reptiles, amphibians, and fish provide an unmistakable view of the types of injuries being inflicted upon the Earth's biological systems. When the IUCN numbers are totaled, about one quarter of all vertebrate species are limited to small and localized populations, or declining steeply in numbers, or already threatened with extinction. The problems afflicting vertebrates are truly global—threatened vertebrates can be found in virtually every ecosystem and country worldwide—but they unfold in complex ways. Humankind is creating a mosaic of impoverished ecological landscapes, and interpreting the patterns correctly requires close scrutiny of the underlying common strands—particularly if we are to pursue alternative paths for conserving nature's diversity.[65]

Regional and country totals of threatened vertebrates

provide a useful starting point for deciphering the mosaic. The number of threatened species in a country or region usually depends upon the total number of species found there overall. Not surprisingly, countries with high species diversity (gained either through location—in the tropics—or through sheer size) have more threatened species. A more refined yardstick for assessing degree of endangerment is the percentage of a country's fauna that is threatened. When the percentages of threatened birds and mammals are calculated (see Tables 7 and 8), nations in a large "Asia-Pacific" triangle bounded by India, Japan, and New Zealand appear in danger of losing proportionately more species sooner than any other part of our planet. China, India, and Southeast Asia combined hold over 2 billion people and were, until mid-1997, the world's fastest-growing economies. Habitat destruction—led most prominently by regional logging industries—has been widespread, and many native mammals and birds must also contend with heavy hunting pressure. In Australia, the biggest threats have come from invasive species such as cats and foxes, and habitat loss linked to more subtle changes—overgrazing by cattle and sheep over large areas, and a decline in traditional burning patterns maintained by aboriginal cultures. Invasive species are also at the forefront of bird declines in New Zealand, although loss of native forests has played a role as well.[66]

The situation appears somewhat less dire in other regions, but Madagascar, Brazil, Mexico, and the United States stand out as countries with elevated rates of mammal or bird endangerment and high numbers of species at risk. It is also important to note that threatened vertebrates are sometimes concentrated in particularly hard-hit regions of countries. Over 75 percent of threatened bird species in Brazil, for instance, live only in the Atlantic rainforest, seasonally deciduous forest, and caatinga thornscrub habitats in the eastern fringe of the country. Of the 12 primate species endemic to the Atlantic rainforest, 11 are threatened, a much higher percentage than among Amazonian primates. In the United States, states that have seen heavy population increases in recent decades, such as California and Florida, also harbor high numbers of threatened vertebrates.[67]

The leading problem that all vertebrate groups face is habitat alteration. A 1994 mapping study by the organization Conservation International estimated that only 27 percent of our planet's habitable land mass (that which is not ice, rock, or blowing sand) remains "undisturbed"—occurring in large blocks of territory at least 100,000 hectares in size without any permanent settlements, roads, intensive agriculture, or heavy grazing. Undisturbed regions are concentrated around the poles, in the boreal forests of Canada and Russia, in the warm deserts of Africa (the Sahara and the Kalahari), and Arabia, and in the cold deserts of central Asia. All of these regions have their own irreplaceable mix of species and ecological communities, but none are considered "hotspots" of biodiversity. The only remaining large,

TABLE 7

Top 10 Countries with the Highest Percentage of Mammal Species Threatened[1,2]

Country	Total Mammal Species (number)	Mammal Species Threatened (percent)
Madagascar	105	44
Philippines	153	32
Indonesia	436	29
Papua New Guinea	214	27
India	316	24
Australia	252	23
China	394	19
Brazil	394	18
Viet Nam	213	18
Malaysia	286	15

[1]Smaller countries with fewer than 100 mammal species are not included in this analysis. [2]The next 10 countries are: Mexico, Ethiopia, Peru, Thailand, South Africa, Kenya, Colombia, Tanzania, Democratic Republic of Congo, United States.

Source: See endnote 66.

TABLE 8

Top 10 Countries with the Highest Percentage of Bird Species Threatened[1,2]

Country	Total Bird Species (number)	Bird Species Threatened (percent)
Philippines	556	15
New Zealand	287	15
Indonesia	1,531	7
China	1,244	7
United States	768	7
Brazil	1,635	6
India	1,219	6
Viet Nam	761	6
Australia	751	6
Japan	583	6

[1]Smaller countries with fewer than 100 bird species are not included in this analysis. [2]The next 10 countries are: Thailand, Malaysia, Colombia, Peru, Myanmar, Argentina, Mexico, Papua New Guinea, Ecuador, and Russia.
Source: See endnote 66.

undisturbed, species-rich landscapes are in the northwest Amazon Basin, and in a few remote corners of central and south-central Africa.[68]

At the other end of the scale, about 36 percent of the Earth's habitable land is "highly disturbed"—either dominated by urban settlements, permanent agriculture, transportation networks, waterworks, and other modifications that support modern society, or used so heavily for grazing that much of the native vegetation has been lost. The regions of our planet that arguably have seen the most complete conversion of native habitat are the fertile lowland river basins and coastal plains of the temperate zone, particularly eastern China, the midwestern United States, and lowland Europe. These, of course, are the world's great "bread baskets." Most of the rest of the temperate and subtropical regions are a mosaic of "partially disturbed" landscapes—areas that are subject to extensive grazing, logging, shifting agriculture, and other disturbances but still maintain some semblance of their native flora and fauna. During this century, such partially disturbed landscapes have become increasingly dominated by secondary habitats—as exemplified by the young stands of Douglas fir and ponderosa pine across the northwestern United States that now grow where centuries-old trees stood, until logging in the region intensified in the early to mid-1900s. Such secondary regrowth is still a forest dominated by the same tree species, but it cannot come close to supporting all the biodiversity of the ancient stands, as the declines of the spotted owl, Pacific salmon runs, and many other species attest.[69]

Many temperate landscapes (particularly in Europe, the Mediterranean Basin and eastern China) have experienced extensive disturbance for centuries. In tropical regions, by contrast, most habitat loss has occurred this century, at an accelerating pace. Most tropical areas of moderate rainfall—such as seasonally dry forests—are already partially disturbed mosaics. This is true as well for foothill or intermediate-elevation habitats with ample rainfall and good soils, such as the central plateau of Costa Rica, the foothills of the Himalaya, and Africa's Great Lakes region, including Rwanda, Burundi, Uganda, and western Kenya. The upper elevations of mountain ranges tend to be more ecologically intact than lower regions, being less populated and harder to access. Wet tropical lowlands have been some of the last areas to be densely colonized by people—which in part is why rainforests generate so much conservation attention today. The eastern and southern Amazon Basin, the Caribbean lowlands of Central America, the dipterocarp forests of Sumatra and Borneo, and the rainforests of Cameroon and Gabon are all currently frontiers for logging, mining, and agricultural settlement. Like mountain areas, semi-arid landscapes generally host lower human population densities, but due to their ecological sensitivity, most of them are partially or heavily disturbed by livestock overgrazing. Much of the Indian subcontinent falls into this category, as do the Middle East and the Sahelian zone of Africa.[70]

Compared to habitat loss, overhunting threatens fewer species directly, but it is even more widespread. Particularly in tropical and subtropical regions, large mammals and reptiles, prized game birds, and rich fish stocks are now gone from many ecological communities or landscapes—from rainforests to coral reefs—that are otherwise relatively intact. Populations of the most sought-after crocodilian in the Amazon Basin, the black caiman, have been overharvested along virtually all Amazonian rivers, far beyond the logging and colonization frontiers eating away at the fringes of the still vast forest. In some cases, landscapes have rebounded from habitat disturbance, but still lack species that never recovered from overhunting. This is the case with the Appalachian forests of the eastern United States, which today only retain two large mammals—white-tailed deer and black bear—where 200 years ago there were also elk, woodland bison, mountain lions, and wolves. And given the important ecological roles typical of large animals, it is reasonable to assume that excessive hunting and fishing are now a significant ingredient in the disruption of many natural communities.[71]

The spread of exotic species is the third major problem facing vertebrates. Invasive species have wrought significant damage on virtually all tropical and subtropical island chains throughout the world, and have also hit freshwater systems hard, particularly in the United States and other temperate regions. To date, most catastrophic collapses of ecological communities—the cichlids of Lake Victoria, the native birds of Pacific islands, and the little-noticed extinction wave of native mammals on Caribbean islands over the past 300 years—have all featured invasive species as a central element.[72]

In addition to the problems that vertebrates face at present, before long they may have to contend with hazards that are not yet so ubiquitous, but have great potential to become so. One is the systemic pollution of biological food webs by synthetic chemicals—pesticides, solvents, and many other compounds widely used in industrialized and industri-alizing countries. Pollution and chemical contamination have triggered some spectacular vertebrate die-offs, and have occasionally brought species—particularly birds of prey—to the brink of extinction, but not on the scale of habitat loss, overhunting, and invasives. This could change, however, if the chronic, cumulative build-up of persistent toxic com-pounds in the environment is allowed to continue.[73]

No corner of the Earth is free from such contamination today. Black-footed albatrosses, masterful soarers that spend their lives far out at sea and breed on tiny remote Pacific atolls, carry DDT, PCBs, and dioxin-like compounds in their tissues in concentra-tions equaling those in bald eagles from the heavily industrialized U.S. Great Lakes region. It is thought that the chem-icals accumulated by the albatrosses orig-inate from plastics burned in municipal incinerators in Japan, and from a plume of toxins that drifts out over the Pacific from Southeast Asia. Synthetic pollutants are also a leading culprit for reproductive failures and grotesque developmental abnormalities suffered by frogs in Europe and the United States. We already know that many of these toxics at their current concentrations in ecological food webs pose serious health risks for people who rely on fish or wild game for a large portion of their diet. If these trends continue, the toll of species losses from chemical contamination could become much more promi-nent.[74]

For most biodiversity, the only ark that offers safe passage is planet Earth.

Another looming problem whose ultimate effects on biodiversity have not yet been felt is global climate change. If the current scientific consensus on the rate and scale of climate change proves accurate, then over the next century natural communities will face a set of unprecedented pres-sures. A warmer climate will probably mean changes in sea-sonal timing, rainfall patterns, ocean currents, and various other parts of the Earth's life support systems. In the evolutionary past, the ecological effects of abrupt climate shifts were somewhat cushioned by the possibility of movement. One part of a plant's or animal's range might dry out, for example, and become inhospitable, but another area might grow more moist and become available for colonization. Today, with more and more species confined only to fragmented remnants of their former range, compensatory migration is less and less likely.[75]

Both current and future pressures on vertebrates illustrate that if our governments and economic systems continue along their present tack, the ecological endowment we bequeath to our children will be greatly diminished. Such a projection for the future, however, assumes that vertebrates will accurately foretell the fate of plants, insects, isopods, corals, polychaete worms, and other non-vertebrate organisms that make up the bulk of the world's species. On the one hand, these groups generally do not have the resource demands of vertebrates—a characteristic likely to reduce their susceptibility to extinction. But plants and invertebrates also tend to be very rich in localized, endemic species with small ranges and specific habitat requirements. These last two factors increase the likelihood that habitat loss or invasive species could pose significant problems and lead to species losses.

Already zoologists have documented major population declines and extinctions among some better-known invertebrate communities. Of the 300 freshwater mussel species and subspecies native to the United States, some 7 percent are already extinct and 40 percent more are gravely threatened by a combination of dams, declining water quality, and exotic species invasions. The same combination of factors in the Mobile River Basin, centered in the state of Alabama, has virtually wiped out what was probably the world's most diverse freshwater snail community. Of some 118 native snail species, 38 are already gone, and 71 of the 80 remaining species are well on their way to the same fate. On Hawaii and at least 20 other Pacific and Indian Ocean archipelagos, the predatory rosy wolfsnail (a native of Central America) and a parasitic flatworm were both intentionally introduced to control invasive snail pests. This poorly planned attempt at biological control is now mowing down an unintended target—the spectacular native snail diversity of Indo-Pacific islands. The wolfsnail and the flatworm are the leading cause behind the extinction and endangerment of more than 250 endemic native snail species.[76]

These and other documented declines strongly suggest that invertebrates and plants are just as susceptible as vertebrates are to habitat loss, invasive species, and related problems. They confirm that the real challenge we face is not just safeguarding vertebrates, but protecting all the dimensions of biological diversity. When many of us think about biodiversity conservation, the first image that comes to mind may well be of zoological parks, aquaria, and botanical gardens—modern-day arks protecting rare animals and plants. In one sense, the image is correct; these institutions are crucial for conservation. Zoos, aquaria, and botanical gardens are the last refuge for many species on the brink of extinction; they are also major environmental education centers, and support ecological restoration by breeding species in captivity that can subsequently be reintroduced to suitable habitat in the wild.

But for most vertebrates and the rest of biodiversity, the only ark that offers safe passage is planet Earth itself. While zoos and reintroduction programs address the immediate problems of how to rescue and restore the most highly threatened species, they need to be complemented by a much larger effort to reduce the underlying pressures we are placing on natural systems. If we continue on our current tack, nature in the future is likely to consist of far simpler and more uniform ecological systems that have lost much of the biodiversity and the value to humankind they once had. But this outcome is not yet guaranteed. There is still room and time to sustain a biodiverse world, and many people are striving to do just that—to better manage protected areas and implement other habitat protection measures, to better enforce international agreements covering biodiversity conservation, and to build a more ecologically sound and socially equitable society.

Protecting Habitat

One of the foremost laws anywhere in the world giving sanctuary to wild species is the Endangered Species Act (ESA) of the United States. Passed in 1973, the ESA gives the federal government broad powers to protect species that are judged by the U.S. Fish and Wildlife Service to be endangered or threatened. Some 900 North American species have received federal protection under the ESA, which entitles a species to receive, among other things, research attention, captive breeding to boost its numbers, and restrictions on human activities in its habitat. Of the 128 species originally listed under the Act in 1973, about 59 percent are currently stable, improving, or—like the American alligator and the brown pelican—have recovered so successfully that they are off the roll entirely. But the species-by-species approach that initially served the ESA well now appears to be breaking down. Certain regions in the United States are under so much pressure from land development that entire natural communities and habitats are unravelling. The ESA review and recovery process cannot keep up with the numbers of declining species—at last count, there were over 4,000 candidate species awaiting final review. Since intensive care for endangered species is relatively expensive to fund, habitat protection is the best long-term investment a country can make for biodiversity.[77]

The first step in safeguarding a particular habitat is understanding how it is being or has been degraded. This involves investigating ecological dynamics—such as the disruption caused by an exotic tree that invades a native ecosystem—and it also requires examining human dynamics. For instance, the conversion of tropical forest to other land uses in Darién Province of Panama is linked to the actions and aims of commercial logging companies, small landowners (both long-time residents and recent immigrants), large landowners, government officials, distant consumers, international development agencies, and even conservationists. Human population growth is one social trend that is often implicated as fueling habitat loss. In Darién—the easternmost and by far the least populated of Panama's provinces—the resident population climbed from 22,000 in 1970 to 55,500 in 1995, during which time forest cover in the province declined from 94 percent to less than 65 percent. Political and socio-economic factors such as the distribution of land and income in a country are perhaps even more important for making sense of habitat loss. The most extensive deforestation in Darién has occurred on lands settled by colonists from central and western Panama, who tend to pursue a subsistence strategy based on raising cattle, which requires large amounts of cleared land. Colonists began to emigrate to Darién in substantial numbers only after the Pan-American highway was extended into the province in 1980, providing the first overland access. One reason why they move to Darién is that much of the best farmland and pasture in their home provinces is concentrated in the hands of large landowners, leaving poorer residents with limited options for making a decent living.[78]

The main approach that national governments have taken to address habitat loss has been to establish systems of national parks, wildlife refuges, forest reserves, marine sanctuaries, and other formally protected areas. Nations have steadily increased the number and extent of their protected areas during this century. At present, about 1 billion hectares of the Earth's surface is officially designated as protected, an area nearly equal to the size of Canada.[79]

Protected lands safeguard some of the world's greatest natural treasures, and have made a big difference for some "conservation-dependent" vertebrates that would otherwise almost certainly be sliding into extinction. In East and southern Africa, for instance, these include about 40 species of the region's famed "megafauna," such as giraffes, hyenas, wildebeest, and impala. The populations of these animals are presently out of danger, in large part because of an extensive reserve network in their home countries. Yet despite these notable successes, current networks of protected areas alone are unlikely to save most biodiversity.[80]

One reason for their limited ability to save species is that protected areas do not always target sites of high biodiversity. Icy mountain peaks, for instance, are obvious and easy sites for national parks due to their spectacular scenery and lack of development pressure, but they are usually not hotspots of species diversity. Although the world added more protected areas—1,431 new reserves totalling 224 million hectares—between 1990 and 1995 than during any previous five-year period, most of this increase came from a few huge designations in lightly populated desert and high mountain areas, such as the empty quarter of Saudi Arabia and the Qiang Tang Plateau in western China. And many highly diverse ecosystems—from tropical wetlands and dry forests to large temperate river basins—continue to receive little formal protection.[81]

To help ensure that future protected area designations do as much as possible to preserve the diversity of life, conservation organizations are putting biodiversity hotspots down on paper. The organization BirdLife International has mapped 218 regions around the globe that contain concentrations of bird species with restricted ranges of less than 50,000 square kilometers. (About one quarter of all bird species fall into this category, including 74 percent of all threatened birds.) These "Endemic Bird Areas"—which originally covered a mere 1 percent of the Earth's land surface but have already lost half their area to habitat conversion—can serve as indicators of regions likely to contain high numbers of endemic species for other less well known taxonomic groups. In a slightly different approach, the World Wildlife Fund and the Nature Conservancy have begun mapping *ecoregions*—geographic areas defined by the unique plant and animal diversity they contain—as a priority for deciding where to target conservation efforts. Ecoregions have recently been mapped on a continent-wide scale for the United States, Latin America, and the Caribbean, and a similar effort is currently under way for the Asia-Pacific region.[82]

Conservationists are also comparing the distribution of ecological communities against protected area networks in what is called a "gap" analysis, looking for communities not represented in existing reserves. Costa Rica, for instance, has one of the most advanced systems of protected areas of any nation, but most reserves are situated either above 1,000 meters elevation or below 50 meters. As a result, foothill forest habitats rich in distinctive species are underrepresented in officially protected areas. These middle-elevation forests are also where most Costa Ricans live and work—and thus where forest fragmentation has been most extensive, which also puts them high on the list for conservation.[83]

Brazilian Amazon park guards must each patrol more than 6,000 square kilometers.

An even greater shortcoming of the reserve approach is a lack of implementation. On paper, national governments have designated about 6 percent, on average, of their respective land and marine territory as protected areas—but in reality, many of these are largely unprotected. These "paper parks" are most common in developing countries, which hold the bulk of the world's biodiversity yet have the least in the way of money or expertise to devote to managing protected areas. Ministries and agencies responsible for protected areas management are chronically underfunded and understaffed in most countries, particularly in the developing world. In Africa, for instance, government budgets for protected areas in the late 1980s averaged less than one fifth of the amount necessary to support even a minimum standard of management. Brazilian park guards in Amazonian protected areas must each cover on average more than 6,000 square kilometers, an area larger than the U.S. state of Delaware. Moreover, in many governments, agencies responsible for natural resource management are politically marginalized and have little bureaucratic influence. This limits their ability to respond to outside threats to protected areas—particularly when those threats are encouraged by other government policies. As a result, many officially designated reserves are subject to logging, agricultural development, mining, extensive poaching, and other forms of degradation.[84]

Such scant commitment to protected areas makes it easy to decommission them with a stroke of a pen—an all-too-frequent consequence of the rush toward some short-term bonanza in natural resource exploitation. In India, for instance, politicians reduced the size of the Melghat Tiger Reserve by one third in 1992 to accommodate timber harvesting and dam construction, while more than 40 percent of the Narayan Sarovar Sanctuary was turned over by the Gujarat State Assembly in 1995 to mining companies eager to harvest the coal, bauxite, and limestone deposits found there. Narayan Sarovar was home to a rich assembly of wildlife, including wolves, desert cats, and the largest known population of the Indian gazelle.[85]

The "paper park" syndrome has deep roots; because in part it reflects a lack of political commitment to the conservation of wildlands and natural habitats, it cannot be cured simply by increased international funding for protected areas management. The tactics for building broad-based constituencies for biodiversity will vary from one society to another, but virtually everywhere the effort will require two basic strategies. Environmental education programs built into school curricula (preferably beginning at the primary level) can help people understand the complexity of natural systems, and the importance of caring wisely for them. And practical, culturally sensitive initiatives are needed to restore local residents' roles in natural resource management, particularly where they have traditionally safeguarded landscapes rich in biodiversity.[86]

Biodiversity-related education can be as straightforward as community elders recounting stories to their grandchildren about the lives of animals or plants in surrounding lands. In Seri Indian villages along the Sea of Cortez in northwest Mexico, schoolteachers, elders, and conservationists have documented traditional Seri songs in a book for childrens' use in community schools. The songs reflect the rich and intricate knowledge that the Seri—traditionally a nomadic hunting and gathering culture—have about the animals, plants, ocean, and desert where they live. School teachers in the United States can now access over 200 curriculum guides, videos, posters, and other educational resources that present biodiversity themes for different grade levels of students. Such materials are generally not available to students in biodiversity-rich developing countries, particularly in rural areas, but there are plenty of other ways to get a conservation message across. In Indonesia, a forestry education program presented conservation themes using traditional shadow puppet theater, while in Uganda, grassroots development organizations have used popular theater and dance performances to tackle environmental topics such as soil erosion and biodiversity loss.[87]

A great deal of the natural wealth that conservationists have sought to protect by means of parks is actually on land and under waters that have long been managed by local people. Long-established native communities throughout Asia, Africa, and the Americas, have traditionally protected forests, mountains, and rivers as sacred sites and ceremonial centers. In some areas of Sierra Leone, for example, the best remaining native forest patches are found within sacred groves maintained by local villages. Throughout Central America, from the Mayan lowlands of Guatemala and Belize to the rainforest of Panama's San Blas coast, most remaining extensively forested areas—as well as some of the most diverse coral reefs and other coastal ecosystems—overlap uncannily with homelands of indigenous cultures. Local people often have a great fund of pragmatic knowledge: they know how the local weather works; they know which organisms produce powerful biochemicals; they know what grows where.[88]

Environmental education and protected areas management in such situations must work both ways: conservationists can often learn a great deal about biodiversity from those
who have lived within it for generations. Cutting such people out of the loop is not a good idea—some of the biggest
mistakes in natural areas conservation have involved the
forcible removal of long-term residents from newly designated protected lands. Relocating such individuals or denying
them access to long-used plant and animal resources has
generated a great deal of ill will towards protected areas
worldwide. In some cases, local people have reacted by pur-
posefully neglecting plant and animal resources that they
had previously managed wisely for generations.[89]

Given the rocky history of conservation efforts implemented from the outside, it is not surprising that the projects that have been successful share an ability to channel
concrete, sustained benefits from protected areas to the people who live in and around them. Zambia has applied this
kind of approach to its "game management areas," or GMAs,
which are inhabited buffer zones surrounding the country's
national parks. Both the parks and the GMAs were originally managed by the Zambian Parks and Wildlife Service with
central government funds, and wildlife in both was decimated by heavy poaching in the 1970s and early 1980s following cutbacks in administrative budgets and staffing. In
response, the Zambian government instituted reforms that
redirected funds generated by the parks and GMAs (princi-
pally through fees for safari hunting) back to management
of the protected areas and to local community authorities,
rather than just to the central government treasury.
Additional reforms allowed villagers to be hired as game
scouts, and gave local leaders—village headmen and regional chiefs—the authority to appoint them.[90]

In the Lupande GMA, where this new approach was
first implemented in 1985, villagers quickly went from
regarding the GMA as an edict imposed by distant bureau-
crats to take wildlife away from them, to seeing wildlife as
something they could once again benefit from—and now

had to protect against outsiders. In addition, village scouts proved to be far more effective in patrolling their GMA than the civil servant scouts; they knew their home terrain better, missed fewer days on the job, and worked closely with local chiefs and other villagers to stop poaching. As a result, local populations of elephants and other wildlife showed signs of recovery by the late 1980s, and Zambia has since extended the reforms to GMAs across the country. Zimbabwe has taken a parallel approach to wildlife management with its CAMPFIRE program, and has achieved similarly positive results.[91]

Since some 70 percent of the world's protected areas are presently inhabited by human communities, safeguarding habitat in the 21st century will continue to be as much a social as a biological endeavor. Since the 1970s, a variety of approaches, such as biosphere reserves and integrated conservation and development projects, have sought to address the social complexity of conservation. Their ultimate success depends on the ability and commitment of conservationists, governments, and local residents to address in good faith concerns about land tenure, community development, access to natural resources, and other thorny issues. This is a tough agenda for anyone to tackle, and it is no wonder that, as forester Jeffrey Sayer notes, conservation and development projects "...have tended to work best when an individual or a small group of people have committed themselves to dealing with the conservation problems of an area over a long period of time, and have done so in [ways] sensitive to the needs and values of local people."[92]

In Central America, most remaining forests overlap with homelands of indigenous cultures.

One initiative that has brought together community and conservationists' concerns is the Annapurna Conservation Area Project (ACAP) in Nepal. Covering 7,600 square kilometers, the Annapurna region is one of the most dramatic stretches of mountain scenery on the planet, with peaks over 8,000 meters high, the world's deepest gorge, and diverse habitats from subtropical forest to alpine grasslands. The region shelters highly threatened snow leopards and lesser pandas (a smaller, red-furred relative of China's giant panda), along with the near-threatened Himalayan musk deer and blue sheep, and five different pheasant species. Annapurna is also home to nearly 120,000 people representing 11 distinct ethnic groups. Over the past several decades, it has become the most popular trekking destination in Nepal—some 43,000 tourists were visiting the region annually by the mid-1990s to hike the footpaths that wind beneath the soaring mountains.[93]

Nepali conservationists realized that Annapurna's ecology and traditional cultures were vulnerable to degradation—declines in forests and water quality from uncontrolled tourism development were evident by the early 1980s. But they also recognized that a strictly protected national park was inappropriate for such a densely populated region. Instead, in 1986, non-governmental organizations active in Nepal recommended an experimental "conservation area," where management would be carried out by village-based conservation and development *panchayats*, or committees.[94]

From the onset, the individuals who charted and—with government approval—implemented ACAP put substantial time into discussing plans with local residents and gaining their approval. ACAP proposed a straightforward partnership: villagers were responsible for managing desired conservation projects (including contributing labor, materials, and a proportion of necessary funds) and for proposing and enforcing regulations to protect forests, wildlife, water supplies, and other natural resources. ACAP workers, in turn, were responsible for "motivating people about the environment," as one staff member put it. This included encouraging the formation of local committees; introducing ideas from forestry, wildlife management, tourism development, and other technical fields; coordinating regional environmental education; and helping implement small-scale projects to provide village hydroelectric generators, tree nurseries, trail maintenance, and other local infrastructure.[95]

In one short decade at Annapurna, village committees have revived traditional management of local forests; they enforce when and where trees may be felled for timber and fuelwood, and manage the revenue derived from forest use fees and fines. Some local regulations—for instance, banning all exploitation of the rarest timber trees—have been more stringent than those promoted by Nepal's national forestry department. Village committees have also taken the lead on wildlife protection, often banning all hunting outright and conducting campaigns to remove snares set for musk deer by outsiders involved in the musk trade. Some villages also have ceased offering long-standing bounties for killing snow leopards—a policy that was originally promoted to protect livestock. Reforestation and clean-up campaigns have been carried out, and local committees have made progress in putting tourism development on a more sustainable footing. Tourist lodges in several parts of Annapurna are now required to cook and heat using kerosene (supplied by an ACAP-sponsored fuel cooperative) and other alternative fuels, which spares the three tons of high-altitude rhododendron and birch forests that were being cleared daily for fuelwood. In one area, seven lodges were dismantled and relocated to protect forest along a trail corridor.[96]

Several factors explain how the Annapurna project was able to catalyze local conservation advances. First, the project was designed to be implemented over 10 years, a longer commitment than most conservation or development agencies usually make. ACAP staff also made an extraordinary effort to explain the project to Annapurna residents, most of whom were extremely suspicious of the organization at first, fearing it would impose an unwanted national park. The staff of ACAP—most of whom were hired locally—proved to be well-trained, dedicated, and skilled communicators. The project also was committed to working on communities' rural development priorities as much as on ecological conservation per se.

Perhaps the most important asset of ACAP, however, was the willingness of senior project staff to grant ultimate decisionmaking authority on all project activities to local villagers. This left staff free to concentrate on developing a strong presence as advocates and catalysts for a sustainable model of conservation and development. Project staff strove to strengthen local cultural traditions that promoted sound natural resource management, but they also advocated change where they felt it necessary. For instance, the project has promoted the role of women and minority ethnic groups on conservation committees, something that probably would not have happened otherwise since local politics and governance issues are dominated by men. Although the Annapurna Conservation Area still has many issues to address and problems to resolve, the conservation process there is advancing.[97]

Although we can point to progress in places like Zambia and Annapurna, there is, of course, no guarantee that all local people will make equally good partners for conserving biodiversity. There can be a big difference between the attitudes of a community whose members have lived in a particular valley for many generations, and the views of a community whose residents recently immigrated to a forest frontier in search of land from which to make a living. Sometimes inequities or corruption in governance systems overwhelm even the best-intentioned protected areas, as can social divisions within local communities or between ethnic groups. Many of these problems have bedeviled the Dzanga-Shanga Special Reserve of the Central African Republic, where the World Wildlife Fund has sponsored a conservation and development project since 1988. The populace of Dzanga-Shanga has mostly immigrated to the region, represents 10 ethnic groups, and does not have well-established land tenure. Unlike Nepal, Dzanga-Shanga has no strong communal management traditions for forest or other natural resources. The conservation project has also had trouble generating economic opportunities to match those offered by bushmeat markets and logging operations in the region. As a result, the project has been unable to foster locally based institutions or social structures capable of limiting exploitation of wildlife and other natural resources within the Dzanga-Shanga region.[98]

Seeking Global Solutions

At its root, the accelerating loss of vertebrate species and other strands in the fabric of life is driven by the unprecedented expansion of humanity and its material demands in the last century. Unless we can develop a new balance between human needs and those of the natural world, no conservation areas or parks, no matter how well supported, will be able to head off further deterioration of biodiversity. We must come to terms with population growth, inequitable social systems, and short-sighted economic practices if we are to restore the ecological health of our planet, particularly in the landscapes and ecosystems we depend upon for food, water, fiber, energy, and our other material needs. In a sense, this is conserving habitat not so much by establishing specific zones or regions for conservation, as by developing more ecologically and socially sustainable ways of managing wildlife populations, raising crops and livestock, securing energy supplies, procuring, processing, and using timber and fibers, and guiding national economies and international investment. Local, national, and international efforts must all be linked in this effort to deal effectively with the underlying pressures on the landscapes and ecosystems that support biological diversity.

According to United Nations population experts, human numbers are likely to climb from their 1995 total of 5.7 billion to between 7.7 and 11.2 billion by the middle of the next century. The population total that prevails will depend especially upon the availability and effectiveness of family planning programs, women's access to education, and other social factors. Many developing countries, where most of the new people on Earth will be born—and which shelter most of our planet's biodiversity as well as most currently threatened species—already experience the strains a steadily expanding population places on land, natural resources, and social systems. At the same time that regional chiefs and chieftainesses in Zambia, for example, are regaining their traditional role in managing wildlife, they are facing the effects of a rural population boom. They are not able to allocate as much fallow farmland to needy families as is customary—most good land is already taken, and people are having to cultivate smaller and more marginal plots. As a result, the local ecology suffers along with the local people. Extending the reach of national family planning programs in Zambia and other African countries could play a decisive role in determining whether the continent in the year 2050 will hold 1.7 billion people (the low-fertility scenario), 2.0 billion (the medium-fertility scenario), or 2.4 billion (the high-fertility scenario). If Africa is able to stabilize its population near the bottom of this wide range, the challenge of meeting the peoples' basic needs, as well as protecting the continent's biological wealth, will be less daunting.[99]

The fact that most population expansion will take place in those countries least able to cope with it highlights the importance of developing more ecologically sound approaches to meeting human needs. The challenge is particularly urgent in rural regions worldwide where natural systems remain the most intact. Sometimes the foundations for ecologically sound development are already present in rural traditions and customs—as witness the advancements made in Nepal once authority for forest and wildlife management was returned to village committees. However, virtually all areas today are in the midst of tremendous change—including greater integration into national economies and a corresponding desire for cash income, rising human population densities, and abandonment of cus-

tomary beliefs and knowledge systems by younger generations. Few land management traditions or local political
institutions will be able to confront such trends and avoid
environmental degradation all on their own.[100]

Although managing biological resources wisely and creating a sustainable society will continue to be the work of
communities and nations, such activities are bolstered by an
international legal process that began to develop early in this
century and has accelerated greatly in the last two decades.
One of the most successful international environmental
treaties is the Convention on International Trade in
Endangered Species (1973), which pro-
vides a powerful legal tool for controlling **The WTO**
international trade in wildlife and plants. **deems trade**
The CITES secretariat can request infor- **measures**
mation from governments about species **implemented**
they are trading, and also can demand
reports from countries that appear to be **to protect**
following questionable practices. CITES **threatened**
also provides clear guidelines for customs
agents and inspectors who monitor **species to be**
wildlife shipments and pursue illegal **illegal.**
smuggling rings. CITES was the mechanism through which countries agreed in
1989 to ban international trade in African elephant ivory,
which for two decades had fueled heavy poaching that
reduced elephant numbers from several million to 500,000
or less. Immediately following the ban, African elephant
poaching appeared to drop substantially in many areas.[101]

But CITES and other international agreements are only
part of the puzzle when it comes to reforming networks as
complex and entrenched as the ivory trade. International
demand for ivory, particularly in East Asia, has remained
strong this decade, while a number of elephant range countries, such as the Democratic Republic of Congo (formerly
Zaire) have experienced political instability and declines in
government antipoaching efforts. As a result, poaching
intensity has crept gradually back upward, and illegal ele-

phant kills are again being reported regularly. The real test in nations' commitments to ensuring a sound future for African elephants will come in 1999, when elephant ivory may once again be legally traded on a limited basis. Delegates at the 1997 CITES conference of the parties agreed that three southern African nations with robust elephant populations—Namibia, Botswana, and Zimbabwe—would be able to trade limited volumes of ivory with Japan starting in 1999, provided they establish an independently verifiable system to closely track the entire trade process. Since the three African countries involved have pledged to reinvest the funds generated by the ivory sales back into wildlife conservation, limited trade could benefit elephants—but only if the monitoring system is rigorous enough to keep illegally harvested ivory out of the trade pipeline.[102]

CITES has on occasion been backed up by sanctions and trade restrictions. In 1994, for instance, the United States, acting under a domestic law called the Pelly Amendment, banned the import of fish and wildlife products from Taiwan because of that nation's continuing trade in tiger body parts and rhino horn—both illegal under CITES. Subsequently, Taiwan stepped up the implementation of its wildlife conservation laws, retrained officers responsible for wildlife enforcement, and developed a new education and public outreach campaign on the endangered status of species used medicinally. As a result, the frequency of rhino and tiger parts in Taiwanese shops fell markedly, and the volume of illegal wildlife shipments seized by customs officials increased. In response, U.S. officials lifted trade sanctions against Taiwan in 1996.[103]

While such unilateral enforcement of wildlife protection laws has gotten results, most countries do not have the political and economic clout to implement unilateral sanctions and standards effectively. These actions are also extremely controversial and are opposed vigorously in many international quarters. The World Trade Organization (WTO), for one, deems trade measures implemented to protect threatened species to be illegal under the General

Agreement on Tariffs and Trades (GATT). During the early 1990s, the United States restricted yellowfin tuna imports from countries whose fleets did not take steps to eliminate tuna-fishing practices known to kill large numbers of dolphins. (Spotted dolphins, the species that swims most often with yellowfin tuna, were estimated in 1993 to have suffered population declines of nearly 75 percent). The U.S. restrictions were appealed by Mexico to a GATT panel, which ruled in Mexico's favor. Subsequently, Mexico agreed to suspend its appeal and entered into direct negotiations with the United States, producing in the end a resolution that allowed the U.S. trade standards promoting dolphin-safe tuna to stand. Nevertheless, in other rulings, the WTO has remained hostile to environmental standards that restrict trade—most recently in April 1998 when it ruled against a U.S. ban on shrimp imports from Asian nations that do not require fishermen to equip their shrimping gear with turtle excluder devices.[104]

Since unilateral sanctions inevitably generate controversy, governments will probably remain reluctant to implement them for environmental reasons unless pushed hard to do so by domestic constituencies. Multilateral sanctions are more difficult to negotiate, but once in place they tend to be harder for countries to avoid through appeals or other means. CITES has the power to call for limited multilateral sanctions, and did so in 1991 on Thailand because of that country's role as a regional center for illegal trade in endangered wildlife. The Thai government did not attempt to fight the sanctions and began enforcing wildlife laws more vigorously.[105]

One area where institutions concerned with international trade could make a positive contribution to biodiversity conservation is in establishing guidelines to control the inadvertent spread of invasive species. Many exotics get to a new place by stowing away in the cargo holds of planes, boats, or trucks, or hitching a ride in the bilge water that ships take in at the start of a voyage and discharge at the end. Some countries already recognize the threat that invasives pose to native species and communities, and monitor incoming ships, planes, and other vessels in an attempt to catch intruders before they become established. The United States since 1991 has required by law close scrutiny of ships and planes that arrive in Hawaii from Guam, the goal being to detect brown tree snakes—which have stowed away in the cargo holds of planes on more than one occasion. Should the snake become established on Hawaii—which like Guam has no equivalent predators—it could similarly devastate the archipelago's remaining native birds.[106]

The most thorough test to date of the international community's will to face up to the biodiversity crisis is the Convention on Biological Diversity (CBD), which came out of the 1992 Earth Summit in Rio de Janeiro. Now ratified by 172 countries, the CBD is a legally binding agreement that commits national governments to reversing the ongoing global decline in biological diversity. Under the convention, countries are supposed to accomplish this feat by actively conserving biodiversity, for instance in systems of national parks, forest reserves, and other protected areas; by making sure biological resources—timber, fisheries, forage for livestock—are used in ecologically sound ways; and by ensuring that the benefits of genetic resources—such as traditional crop varieties used in breeding new lines—are shared fairly and equitably. To implement this approach, governments are required to take a number of steps forward. (See Table 9.)[107]

The CBD is far more comprehensive than other environmental treaties, and the report card on governments' compliance so far has been barely passing. The United States, for one, has not even ratified the convention. The CBD requires each signatory government to prepare a national strategy for conserving their country's biological endowment; the first round of these reports was due in January 1998, but less than half the signatory governments met that deadline. Between 1991 and 1997, donor countries pledged over $3.3 billion in funding to the Global Environment Facility (GEF), which was established in 1991 as a joint initiative of the United Nations Development

TABLE 9

Requirements for Signatory Nations to the Convention on Biological Diversity

- Adopt national biodiversity strategies and action plans.
- Establish nation-wide systems of protected areas.
- Adopt measures that provide incentives to promote conservation and sustainable use of biological resources.
- Restore degraded habitats.
- Conserve threatened species and ecosystems.
- Minimize or avoid adverse impacts on biodiversity from the use of biological resources.
- Respect, preserve, and maintain knowledge, innovations, and practices of local and indigenous communities.
- Ensure the safe use and application of biotechnology products.

Source: See endnote 107.

Programme (UNDP), United Nations Environment Programme (UNEP), and the World Bank to address global environment problems, and was subsequently designated the interim funding mechanism for the CBD. Thirty-eight percent of the GEF's funding to date has been allocated in support of biodiversity conservation, in 156 separate projects. But even with the GEF, wealthy countries have still failed to meet their treaty-specified responsibility to provide "new and additional" financing for conservation. Total annual international aid for programs related to biodiversity conservation (including programs for reducing population growth, combatting deforestation, and promoting women's role in development, sustainable agriculture, and wildlands management) actually fell 11 percent, from $6.4 billion in 1987 to $5.7 billion in 1994.[108]

There is also reason to question the effectiveness of conservation funds spent by the GEF. The institution was restructured in 1994 following an independent evaluation that concluded the GEF spent much of its money on haphazard, poorly executed projects that likely made only a "marginal contribution" to biodiversity conservation. Interestingly, the area of the GEF that appears most promising is the one with the least money to offer—the small-grants program, which provides grants up to $50,000. The program has funded such projects as coastal zone management in Turkey, and a gene bank in the Philippines that preserves native plants with medicinal value; it has been successful enough to receive additional aid from Denmark, Canada, and other countries.[109]

Although the slow pace of implementation reflects the ambivalence of governments about how strong they want the CBD to be, it nonetheless holds promise as an international mechanism for self-policing. If fully implemented, the CBD could also serve as an important forum for minimizing the impacts on biodiversity caused by an an equally international phenomenon—the rapid growth and globalization of economies everywhere. Increasingly long economic chains link natural resources in remote and often biodiversity-rich regions of countries to consumer demands in distant cities and other nations. In some cases, these external economic relationships trigger environmental degradation far more severe than that produced by internal trends like regional population growth.

Some of the most ecologically harmful activities of this kind involve businesses and individuals—both unscrupulous operators and legitimate business ventures—that seek quick exploitation of timber, fish stocks, minerals, and other elements of our planet's natural wealth without regard for the environmental and social damage they inflict. There is often a fundamental imbalance between the political power of rural communities and that of outsiders who have other plans for their lands and resources. Instances where heedless exploitation is occurring are legion: foreign boats rake in fish and lobsters off Nicaragua's Caribbean coast; Indonesian, Malaysian, and French timber companies cruise the rainforests of Gabon and Cameroon; commercial shrimp farms pollute coastal waters and destroy local fisheries in India and Honduras; miners lace Amazonian river beds with toxic mercury compounds to precipitate out gold deposits. Some of these undertakings are sponsored by multinational corporations, which in recent years have found an increasingly friendly international climate in which to operate as countries relax trade restrictions and open their borders. Others primarily benefit developing-country elites who own resource extraction companies—which often garner subsidies and other incentives from their governments.[110]

One reason natural resource extraction continues apace is that the developing countries shoulder heavy loads of international debt. The quick foreign currency generation promised by logging operations, mining companies, or factory fishing fleets is understandably enticing, and financially pressed governments commonly respond with generous concessions to forests, mineral deposits, and fish stocks. All too often, these concessions are granted on remaining natural resource frontiers, where intact or less-disturbed landscapes still hold a wealth of biological diversity.[111]

The onerous debt burdens of developing countries reduce opportunities for conservation in other ways too. Governments that seek to restructure their obligations with international creditors are invariably required to adopt the "structural adjustment" programs of the International Monetary Fund (IMF). While the IMF's intent is to put countries' financial houses in order, the restrictions it insists upon can severely constrain the ability of governments to fund and implement the public policy structure—from environmental lawyers to forest guards—necessary for sound natural resource management at the national level.[112]

Other means of addressing the debt crisis have been more environmentally beneficial. Since the 1980s, a number of debt forgiveness programs have involved "debt for nature" swaps, where governments or conservation groups buy back or forgive a portion of a country's debt (usually at a discounted market price) in exchange for the country government's commitment to fund conservation programs at a specified level in local currency. Debt-for-nature swaps have been concluded in 16 countries and have generated nearly

$130 million in local conservation funds. In some cases, these monies have been used to establish conservation endowments, which then provide a steady source of funds to bolster natural areas management or other programs.

Overall, however, neither debt-for-nature swaps nor other reduction approaches have made a dent in the total indebtedness of developing countries, which has continued to grow steadily, surpassing $2 trillion in 1995. Clearly, debt relief must be undertaken on a much larger scale if affected governments are to dig out from under their overall debt burdens.[113]

Another approach to stemming biodiversity-depleting resource exploitation is to support the development of alternative, more environmentally benign ways for people to make their livelihoods. Ecotourism—or travel oriented around natural sites, native species, and traditional cultural practices—is one commonly proffered alternative economic activity that can extract value from ecological landscapes with minimal harm. Ecotourism has helped spur communities to protect coral reefs near Phuket Island in Thailand and around the Caribbean island of Bonaire. It is also the leading foreign exchange earner in Costa Rica, and helps support that country's protected areas system. "Biodiversity prospecting"—the search for species that might yield new chemicals, drug precursors, or genes—is another commonly touted alternative, particularly as an incentive for protecting species-rich natural communities like coral reefs and tropical forests. A number of bioprospecting agreements between pharmaceutical companies and species-rich tropical countries have been signed in recent years: Merck is working with Costa Rica's National Biodiversity Institute (INBIO), Glaxo with Ghana, Novo is active in Nigeria, and Bristol Myers Squibb is collaborating in Suriname with a consortium of conservationists, indigenous healers, and a Surinamese pharmaceutical company.[114]

Ecotourism, and bioprospecting agreements to a lesser degree, have been effective at generating foreign currency for developing countries, and they also influence national-level environmental policies. The benefits provided to local communities and to actual conservation projects, however, have been much more limited. Realizing these latter benefits depends on how ecotourist and bioprospecting ventures are structured and administered, and who participates in them. If bioprospectors are to make use of flora or fauna from lands traditionally managed by indigenous cultures, they should negotiate agreements with indigenous representatives as well as national government agencies. Clear guidelines need to be in place to ensure that the companies benefitting from bioprospecting provide fair compensation for the right to review samples of species with pharmaceutical potential, as well as royalties on any commercial products that result. Receipts from tourist visits should go to local conservation or community development funds, and not just to tour operators or national treasuries. Local residents should also have a say (as many aboriginal communities in Australia now do) in determining when, where, and how many tourists can visit protected areas—say, for instance, a national park that also contains sites sacred to residents' religious beliefs.[115]

Dept-for-nature swaps have generated nearly $130 million in local conservation funds.

Unfortunately, it is all too easy for governments and economic elites to support efforts like ecotourism with one hand, but continue encouraging shortsighted natural resource management with the other. This situation will only change with fundamental government policy reform in many areas—for instance in eliminating subsidies for cattle ranching that involves clearing native forests (as Brazil did in the late 1980s), and rewriting national land and marine tenure laws to uphold the claims of indigenous peoples with strong traditional ties to land and sea resources. While increased funding for sound natural resource management to complement such reforms is certainly needed, this is more likely to make a difference when coupled with reductions in existing subsidies to frontier-style economic activities that damage biodiversity. During the Convention on Biological Diversity's first three years of implementation, for example, donor countries provided about $18 billion in international aid for conservation-related purposes. Yet during that same time period, global subsidies to biodiversity-damaging activities such as fisheries overexploitation, road building, and excessive fossil fuel burning totaled approximately $1.8 trillion—100 times as much.[116]

It is also important to realize that national resource exploiters are doing the bidding of consumers—especially those of us in developed countries and the rapidly industrializing nations of the Pacific Rim. Take the massive fires that swept across vast areas of the Indonesian islands of Kalimantan and Sumatra during 1997, destroying over 100,000 hectares of primary tropical forest and peat swamp forest—critical habitat for endangered creatures like Asian elephants and orangutans—and enfolding 70 million people in a smoky, sickening haze. These fires were started deliberately, mostly by 176 companies and concessionaires who took advantage of the unusually dry El Niño conditions to clear land for industrial pulpwood and oil palm plantations. Oil palm, which yields an edible vegetable oil, has been heavily promoted as a plantation cash crop by the Indonesian government—the country's palm oil exports rose 162 percent between 1987 and 1997, from 986,000 tons to 2.58 million tons. Recently, the Suharto government has seen agribusiness development as a primary way for Indonesia to dig itself out of its current economic crisis, and oil palm plantations are scheduled to expand even more frenetically, from 2 million hectares to over 3 million within the next two years.[117]

Part of Indonesia's increased pulpwood production will be destined for industrialized countries such as Japan and the United States, while the world's fastest-growing major market for Indonesia's palm oil is China. China's annual consumption of oilseeds (of which oil palm is one kind) has risen steadily from 31.6 million tons in 1987 to 43 million tons in 1997, the result of population increases as well as economic growth permitting more people to eat higher on the food chain. China cannot meet its increased demand for vegetable oil with domestic production alone, and thus has turned to international markets. Annual Chinese imports of palm oil have risen 316 percent during the last 10 years and now top 1.5 million tons. This chain of events illustrates how the ecological price for growth in today's global economy—whether in North America, Europe, Asia, or other emerging markets—is often not paid by those who reap its benefits. And it is not just biodiversity that suffers; replacing native forests with plantations in Indonesia has meant that local subsistence farmers now have fewer sources of the wild forest foods they traditionally rely upon during lean harvest years—and this year's El Niño-afflicted rice crop in Kalimantan is expected to be very poor indeed.[118]

One route for inserting more ecological and social considerations into the national and international chains that supply consumer goods is programs that certify products which meet high environmental and social standards. Environmental certification programs have been implemented most extensively for sustainably harvested timber, primarily through the efforts of the Forest Stewardship Council, a body established in 1993 by environmental groups, indigenous peoples representatives, and progressive foresters and businesses to establish uniform guidelines for certification. By 1997, more than 6 million hectares of forest land had been certified as being under ecologically sound timber management. This is still only a small fraction of the world's forest estate, but the number of certified timber-harvesting operations is growing rapidly. A consortium of the World Bank and environmental and business representatives recently endorsed a resolution to increase the forest area under certified management to 200 million hectares within 10 years. Most of the demand for certified timber has come from the United States and Europe—in Great Britain, for instance, a buyer's group of 75 companies comprising 25 percent of the nation's wood demand has pledged to phase out all purchases of non-certified wood products. So far, little demand for certified timber has materialized in Asia, although the concept has been introduced to Japan, the world's largest importer of whole logs.[119]

Certification programs can also provide specific benefits for conservation of our planet's biodiversity endowment. In the Americas, many migratory songbirds that breed in North America spend part of their winters in coffee plantations from Mexico to Colombia, where coffee bushes have traditionally been grown under a canopy of native forest trees. Unfortunately this habitat is disappearing as plantations intensify and replant with higher-yielding, sun-tolerant coffee varieties that do not require shade—leaving neotropical migrants to search ever harder for suitable wintering territory. Recognition of the importance of traditional shaded coffee plantations for migratory birds and other wildlife has led the Rainforest Alliance to establish a certification program for Latin American shade coffee plantations. Participating farms must maintain forest cover over coffee plants and meet a series of environmental and social criteria for sound management, such as limiting agrochemical applications, controlling soil erosion, and providing their workers with fair wages and environmental education training. Certified shade-grown coffee from Guatemala is now available in the United States through 11 coffee distributors and mail-order retailers, and vendors who sell it emphasize its environmental advantages over standard coffee.[120]

All of these approaches demonstrate that in today's increasingly crowded and interconnected world, the most important steps we can all take to conserve biodiversity may be the least direct ones. The fate of birds, mammals, frogs, fish, and all the rest of biodiversity depends not so much on what happens in parks but what happens where we live, work, and obtain the wherewithal for our daily lives. To give biodiversity and wildlands breathing space, we must reduce the size of our own imprint on the planet. That means stabilizing the human population. It means far greater efficiency in our materials and energy use, and careful consideration of the social and ecological side effects of our international trading links. It means intelligently designed communities. And it means educational standards that build an awareness of our responsibility in managing 3.2 billion years' worth of biological wealth. Ultimately it means creating a less materialist, more environmentally literate way of life.[121]

Humans, after all, are not dinosaurs. We can change. Even in the midst of the current mass extinction, we still largely control our destiny, but we are unwise to delay taking action. The fate of untold numbers of species depends on it. And so does the fate of our children, in ways we can barely begin to conceive.

Notes

1. Western U.S. frog declines from Charles A. Drost and Gary M. Fellers, "Collapse of a Regional Frog Fauna in the Yosemite Area of the California Sierra Nevada, USA," *Conservation Biology*, April 1996; introduced predators as cause from Robert N. Fisher and H. Bradley Shaffer, "The Decline of Amphibians in California's Great Central Valley," *Conservation Biology*, October 1996; agrochemical effects from Oliver Klaffke, "Frog Song—Are Agricultural Chemicals Poisoning Amphibians?" *New Scientist*, 14 February 1998, and from Catherine Baden-Daintree, "Declines Due to Pesticide," *Oryx*, April 1997; UV radiation as cause from Andrew R. Blaustein and David B. Wake, "The Puzzle of Declining Amphibian Populations," *Scientific American*, April 1995; tropical montane frog losses from lecture by Karen Lips, "The Case of the Disappearing Frogs," Smithsonian Tropical Research Institute, Balboa, Panama, 6 January 1998.

2. Australian totals from Jonathan Baillie and Brian Groombridge, eds., *1996 IUCN Red List of Threatened Animals* (Gland, Switzerland: World Conservation Union (IUCN), 1996); Lake Victoria losses from Les Kauffman and Peter Ochumba, "Evolutionary and Conservation Biology of Cichlid Fishes as Revealed by Faunal Remnants in Northern Lake Victoria," *Conservation Biology*, September 1993; Pacific bird losses from David W. Steadman, "Prehistoric Extinctions of Pacific Island Birds: Biodiversity Meets Zooarcheology," *Science*, 24 February 1995; Guam losses from Gordon H. Rodda, Thomas H. Fritts, and David Chiszar, "The Disappearance of Guam's Wildlife," *Bioscience*, October 1997.

3. Baillie and Groombridge, op. cit. note 2; IUCN is the acronym for the original name (the International Union for the Conservation of Nature) for the World Conservation Union.

4. One quarter figure for all vertebrates calculated by Worldwatch from Baillie and Groombridge, op. cit. note 2; this figure is calculated using the total number of vertebrate species assessed by IUCN (rather than total species known), and includes both threatened and near-threatened species.

5. Edward O. Wilson, *The Diversity of Life* (Cambridge, MA: Belknap Press, 1992).

6. Calculations of background extinction rates from David M. Raup, "A Kill Curve for Phanerozoic Marine Species," *Paleobiology*, vol. 17, no. 1, 1991. Raup's exact estimate is one species extinct every four years, based on a pool of 1 million species; the author lists a range of one to 10 species per year based on current estimates of total species worldwide. For Quaternary extinctions, see Paul Martin and Richard Klein, eds., *Quaternary Extinctions: A Prehistoric Revolution* (Tucson, AZ: Academic Press, 1984); estimates for current rates of extinction are reviewed by Nigel Stork, "Measuring Global Biodiversity and Its Decline," in Marjorie L. Reaka-Kudla, Don E. Wilson, and Edward O. Wilson, eds., *Biodiversity II: Understanding and Protecting Our Biological Resources* (Washington, DC: Joseph Henry Press, 1997), and by Stuart L. Pimm, Gareth J. Russell, John L. Gittleman, and Thomas M. Brooks, "The Future of Biodiversity," *Science*, 21 July 1995. To translate rates into whole numbers, we have assumed a total pool of 10 million species.

7.   Importance of wild organisms in pharmaceutical and health care systems from Norman R. Farnsworth, "Screening Plants for New Medicines," in E.O. Wilson, ed., *Biodiversity*, (Washington, DC: National Academy Press, 1988); global over-the-counter drug value from Kenton R. Miller and Laura Tangley, *Trees of Life: Saving Tropical Forests and Their Biological Wealth* (Boston, MA: Beacon Press, 1991); rice example from Robert Prescott-Allen and Christine Prescott-Allen, "Park Your Genes: Protected Areas As *In Situ* Genebanks for the Maintenance of Wild Genetic Resources," in Jeffrey A. McNeely and Kenton R. Miller, eds., *Proceedings from the 1982 Bali World Congress on National Parks* (Washington, DC: Smithsonian Institution Press, 1984); pollination services from Stephen Buchmann and Gary P. Nabhan, *The Forgotten Pollinators* (Washington DC: Island Press, 1996); other natural services from Janet N. Abramovitz, "Valuing Nature's Services," in Lester R. Brown, Christopher Flavin, and Hilary French, *State of the World 1997* (New York: W.W. Norton and Company, 1997), and from David Pimentel et al., "Economic and Environmental Benefits of Biodiversity," *Bioscience*, December 1997.

8.   Estimates for total species on Earth are reviewed by Stork and by Pimm et al., op. cit. note 6; the high percentage of beetles among currently described species is from Terry L. Erwin, "Biodiversity at Its Utmost: Tropical Forest Beetles," in Reaka-Kudla, Wilson, and Wilson, op. cit. note 6.

9.   Stuart L. Pimm and John L. Gittleman, "Biological Diversity: Where Is It?" *Science*, 21 February 1992.

10.   Rachel Carson, *Silent Spring* (Boston: Houghton Mifflin, 1962); Robert J. Hesselberg and John E. Gannon, "Contaminant Trends in Great Lakes Fish," in Edward T. LaRoe et al., eds., *Our Living Resources: A Report to the Nation on the Distribution, Abundance, and Health of U.S. Plants, Animals and Ecosystems* (Washington, DC: National Biological Service, 1995).

11.   Baillie and Groombridge, op. cit. note 2; comments by Edward O. Wilson from roundtable discussion at Williams College, Williamstown, MA, January 1990.

12.   Dates for bird conservation assessment from Baillie and Groombridge, op. cit. note 2. Bird species total rounded off from ibid., and from Howard Youth, "Flying Into Trouble," *World Watch*, January/February 1994.

13.   Threatened species statistics from Baillie and Groombridge, op. cit. note 2; information on crested ibis from James A. Hancock, James A. Kushlan, and M. Philip Kahl, *Storks, Ibises and Spoonbills of the World*

(London: Academic Press, 1992); and from "Species Under Threat—Crested Ibis," <http://www.wcmc.org.uk/species/data/species_sheets/crestedi.html>, viewed 10 January 1998.

14. Ralph Costa and Joan Walker, "Red-Cockaded Woodpeckers," in LaRoe et al., op. cit. note 10.

15. Baillie and Groombridge, op. cit. note 2.

16. Percentage of threatened birds facing habitat loss calculated by Worldwatch from Baillie and Groombridge, op. cit. note 2; Pará logging from Christopher Uhl et al., "Natural Resource Management in the Brazilian Amazon," *Bioscience*, March 1997, and Virgílio M. Viana, "Certification as a Catalyst for Change in Tropical Forest Management," in Virgílio M. Viana et al., eds., *Certification of Forest Products* (Washington, DC: Island Press, 1995); Three Gorges Dam from International Rivers Network, "Three Gorges Campaign," <http://www.irn.org/programs/3g/index.html>, viewed 2 April 1998.

17. Percentage calculated by Worldwatch from N.J. Collar, et al., *Threatened Birds of the Americas* (Cambridge, U.K.: International Council for Bird Preservation, 1992); see also Alison J. Stattersfield, Michael J. Crosby, Adrian J. Long, and David C. Wege, *Endemic Bird Areas of the World: Priorities for Biodiversity Conservation* (Cambridge, U.K.: BirdLife International, 1998).

18. Baillie and Groombridge, op. cit. note 2; one third figure calculated by Worldwatch from IUCN data. Note that certain large islands relatively near mainland areas have bird fauna that are more continental than insular in composition. For clarity, this estimate thus excludes the larger islands of Japan, Sumatra, Borneo, New Guinea, and Cuba.

19. Steadman, op. cit. note 2; Richard Cassels, "Faunal Extinction and Prehistoric Man in New Zealand and the Pacific Islands," and Storrs Olson and Helen James, "The Role of Polynesians in the Extinction of the Avifauna of the Hawaiian Islands," both in Martin and Klein, op. cit. note 6.

20. Baillie and Groombridge, op. cit. note 2.

21. Current endangerment rate from Baillie and Groombridge, op. cit. note 2; Other numbers on Hawaiian birds from Storrs L. Olson and Helen F. James, "Description of Thirty-two New Species of Birds from the Hawaiian Islands," *Ornithological Monographs*, vols. 45 and 46 (Washington, DC: American Ornithologists' Union, 1991), and Leonard Freed, Sheila Conant, and Robert Fleischer, "Evolutionary Ecology and Radiation of Hawaiian Passerine Birds," *Trends in Ecology and Evolution*, July 1987.

22. North American neotropical migrant figures from Bruce J. Peterjohn, John R. Sauer, and Chandler S. Robbins, "Population Trends from the North American Breeding Bird Survey," in Thomas E. Martin and Deborah M.

Finch, eds., *Ecology and Management of Neotropical Migratory Birds* (Oxford, U.K.: Oxford University Press, 1995); European information from Katrin Böhning-Gaese and Hans-Gunther Bauer, "Changes in Species Abundance, Distribution, and Diversity in a Central European Bird Community," *Conservation Biology*, February 1996.

23. Graham M. Tucker and Melanie F. Heath, *Birds in Europe: Their Conservation Status*, BirdLife Conservation Series No. 3 (Cambridge, U.K.: Birdlife International, 1994); Oliver Tickell, "Paradise Postponed," *New Scientist*, 17 January 1998.

24. Ibid.

25. Mediterranean hunting figures from Youth, op. cit. note 12.

26. Information on 1995-96 hawk kills from Les Line, "Lethal Migration," *Audubon*, September-October 1996; update on 1996–97 Swainson's hawk situation from Catherine Baden-Daintree, "Pesticide Withdrawn to Save Hawk," *Oryx*, July 1997.

27. Baillie and Groombridge, op. cit. note 2.

28. Musk deer figure from "Musk Deer Declining Further," *Oryx*, January 1995; other figures from Baillie and Groombridge, op. cit. note 2; Red colobus example from John F. Oates, *African Primates* (Gland, Switzerland: World Conservation Union, 1996).

29. All figures from Baillie and Groombridge, op. cit. note 2.

30. Figure for habitat loss from Baillie and Groombridge, op. cit. note 2; figure for 70 percent endemic primate endangerment calculated from species totals for Asia from A.A. Eudey, *Action Plan for Asian Primate Conservation 1987–1991* (Gland, Switzerland: IUCN, 1987); for Madagascar from Russell A. Mittermeier et al., *Lemurs of Madagascar: An Action Plan for Their Conservation 1993–1999* (Gland, Switzerland: IUCN, 1993); for Atlantic forest from Anthony B. Rylands, Russell A. Mittermeier, and Ernesto Rodriguez Luna, "A Species List for the New World Primates (Platyrrhini): Distribution by Country, Endemism, and Conservation Status According to the Mace-Lande System," *Neotropical Primates*, September 1995, and from threatened status of primates as per Baillie and Groombridge, op. cit. note 2.

31. European cetacean status from Mark Simmonds, "Saving Europe's Dolphins," *Oryx*, October 1994; Baltic seal information from M. Olsson and A. Bergman, "A New Persistent Contaminant Detected in Baltic Wildlife: Bis (4-Chlorophenyl) Sulfate," *Ambio*, March 1995; general information from Thomas Jefferson, Stephen Leatherwood, and Marc Webber, *Marine Mammals of the World* (Rome: United Nations Environmental Programme/Food and Agricultural Organization (UNEP/FAO), 1995).

32. Hunting as endangerment figure calculated by Worldwatch from Baillie and Groombridge, op. cit. note 2.

33. Annual Amazon Basin hunting estimate and neotropical defaunation from Kent H. Redford, "The Empty Forest," *Bioscience*, June 1992; neotropical defaunation also discussed in E.F. Raez-Luna, "Hunting Large Primates and Conservation of the Neotropical Rain Forests," *Oryx*, January 1995; primate taboos from Eudey and from Mittermeier et al., op. cit. note 30.

34. Kim Hill, in consultation with Tito Tikuarangi, "The Mbaracayú Reserve and the Aché of Paraguay," in Kent H. Redford and Jane A. Mansour, eds., *Traditional Peoples and Biodiversity Conservation in Large Tropical Landscapes* (Arlington, VA: Nature Conservancy, 1996).

35. Comparison of subsistence and market hunting from Sally A. Lahm, "Utilization of Forest Resources and Local Variation of Wildlife Populations in Northeastern Gabon," in C.M. Hladik et al., eds., *Tropical Forests, People and Food*, volume 13, Man and the Biosphere Series (Paris: United Nations Educational, Scientific, and Cultural Organization (UNESCO), 1993); and from R.E. Bodmer, T.G. Fang, and L.M. Ibanez, "Ungulate Management and Conservation in the Peruvian Amazon," *Biological Conservation*, vol. 45, 1988; bushmeat as main rural income source from David S. Wilkie, John G. Sidle, and Georges C. Boundzanga, "Mechanized Logging, Market Hunting, and a Bank Loan in Congo," *Conservation Biology*, December 1992; figure on Gabonese bushmeat consumption from Michael McRae, "Road Kill in Cameroon," *Natural History*, February 1997.

36. Peter Jackson and Elizabeth Kemf, *Tigers in the Wild: A WWF Status Report* (Gland, Switzerland: World Wide Fund for Nature, 1996); Peter Matheissen, "The Last Wild Tigers," *Audubon*, March-April 1997.

37. Ecological role of large mammals from John W. Terborgh, "The Big Things That Run the World—A Sequel to E.O. Wilson," *Conservation Biology*, December 1988; elephant-dispersed trees from R.F.W. Barnes, "The Conflict between Humans and Elephants in the Central African Forests," *Mammal Review*, vol. 26, no. 2/3, 1996; whales and deep-sea biodiversity from Cheryl Butman, James T. Carlton, and Stephen Palumbi, "Whaling Effects on Deep-Sea Biodiversity," *Conservation Biology*, April 1995.

38. Jim Robbins, "In 2 Years, Wolves Reshaped Yellowstone," *The New York Times*, 30 December 1997.

39. Ibid.

40. Figure for total Australian mammal extinctions from Ross McPhee and Clare Flemming, "Brown-eyed, Milk-giving...and Extinct," *Natural History*, April 1997; percentage of current fauna threatened and other details of Australian mammal declines from Jeff Short and Andrew Smith, "Mammal Decline and Recovery in Australia," *Journal of Mammalogy*, vol.

75, no. 2, 1994, and from Michael Common and Tony Norton, "Biodiversity: Its Conservation in Australia," *Ambio*, May 1992.

41. Reptile species total from Harold G. Cogger, *Reptiles and Amphibians of Australia* (Ithaca, NY: Reed Books/Cornell University Press, 1992); amphibian species total from Darrel R. Frost, ed., *Amphibian Species of the World: A Taxonomic and Geographic Reference* (Lawrence, KS: Allen Press/Association of Systematics Collections, 1985); assessment percentages and assessment status of reptile groups from Baillie and Groombridge, op. cit. note 2.

42. Threatened herpetofauna figures from Baillie and Groombridge, op. cit. note 2; Brazil as leading in amphibian diversity calculated from Frost, op. cit. note 41; Mexico as leading in reptile diversity from Jeffrey A. McNeely et al., *Conserving The World's Biodiversity* (Washington, DC, and Gland, Switzerland: World Resources Institute, IUCN, World Bank, World Wildlife Fund, and Conservation International, 1990).

43. Habitat loss percentage calculated from Baillie and Groombridge, op. cit. note 2; Galápagos Islands information from Stephen Herrero, "Galápagos Tortoises Threatened," *Conservation Biology*, April 1997.

44. Baillie and Groombridge, op. cit. note 2.

45. Status of all seven turtle species as endangered from Baillie and Groombridge, op. cit. note 2; general information on sea turtles' problems from Howard Youth, "Neglected Elders," *World Watch*, September/October 1997; Center for Marine Conservation, "Sea Turtles," <http://www.cmc-ocean.org/2221m1_seaturtle.html>, viewed 15 April 1998.

46. Southeast Asia tortoise and river turtle trade information from Catherine Baden-Daintree, "Threats to Tortoises and Freshwater Turtles," *Oryx*, April 1996; CITES information from Convention on International Trade in Endangered Species of Wild Fauna and Flora, "Text of the Convention," <http://www.wcmc.org.uk/CITES/english/text.html>, viewed 15 April 1998.

47. Information on Amazonian crocodilians from P. Brazaitis et al., "Threats to Brazilian Crocodilian Populations," *Oryx*, October 1996; other information on crocodilians' status from Youth, op. cit. note 45.

48. Habitat loss figure for amphibians from Baillie and Groombridge, op. cit. note 2; loss of wetlands as a problem for amphibians from Blaustein and Wake, op. cit. note 1; impacts of roads and traffic from "Frogs and Toads Take the Road Less Traveled," *Delta* (Journal of the Canadian Global Change Program), vol. 5, no. 3, fall 1994.

49. Western U.S. frog declines from Drost and Fellers, op. cit. note 1; Australian frog declines from William F. Laurance, Keith R. McDonald, and Richard Speare, "Epidemic Disease and the Catastrophic Decline of

Australian Rain Forest Frogs," *Conservation Biology*, April 1996; UV radiation and synergistic combinations from Blaustein and Wake, op. cit. note 1; introduced predators as cause from Fisher and Shaffer, op. cit. note 1; drought as problem from J. Alan Pounds and Martha L. Crump, "Amphibian Declines and Climatic Disturbance: The Case of the Golden Toad and the Harlequin Frog," *Conservation Biology*, March 1994; agro-chemicals from Klaffke and from Baden-Daintree, op. cit. note 1; fungal infections and synergistic combinations from Lips, op. cit. note 1; synergistic combinations from Blaustein and Wake, op. cit. note 1.

50. Lips, op. cit. note 1.

51. Baillie and Groombridge, op. cit. note 2.

52. Ibid.

53. Large rivers as hotspots for endangered fish and saltwater hotspots from Peter B. Moyle and Robert A. Leidy, "Loss of Biodiversity in Aquatic Ecosystems: Evidence from Fish Faunas," in P.L. Fiedler and S.K. Jain, eds., *Conservation Biology: The Theory and Practice of Nature Conservation, Preservation, and Management* (New York: Chapman and Hall, 1992); tropical peat swamps as fish diversity hotspots from Peter K.L. Ng, "Peat Swamp Fishes of Southeast Asia: Diversity Under Threat," *Wallaceana*, vol. 73, 1994.

54. Figure for dams worldwide from Sandra Postel, *Last Oasis* rev. ed. (New York: W.W. Norton and Company, 1997); Mississippi dead zone from Joby Warrick, "'Dead Zone' Plagues Gulf Fishermen," *Washington Post*, 24 August 1997.

55. Habitat endangerment percentage and darter total calculated from Baillie and Groombridge, op. cit. note 2; Sanjay Kumar, "Indian Dams Will Drive Out Rare Animals...While Fish Fall Prey to Progress," *New Scientist*, 4 February 1995.

56. Wayne C. Starnes, "Colorado River Basin Fishes," in LaRoe et al., op. cit. note 10; Salvador Contreras and M. Lourdes Lozano, "Water, Endangered Fishes and Development Perspectives in Arid Lands of Mexico," *Conservation Biology*, June 1994.

57. Don Hinrichsen, "Coral Reefs in Crisis," *Bioscience*, October 1997.

58. Ibid.; Fiona Holland, "Reef Wreckers," *New Scientist*, 25 October 1997; Anne Platt McGinn, "Promoting Sustainable Fisheries," in Lester R. Brown, Christopher Flavin, and Hilary French, *State of the World 1998* (New York: W. W. Norton, 1998).

59. Percentage of fish threatened by invasives from Baillie and Groombridge, op. cit. note 2; Richard Ogutu-Ohwayo, "Nile Perch in Lake Victoria: Effect on Fish Species Diversity, Ecosystem Functions and

Fisheries," in O.T. Sandlund, P.J. Schei, and A. Viken, eds., *Proceedings of the Norway/U.N. Conference on Alien Species* (Trondheim, Norway: Directorate for Nature Management and Norwegian Institute for Native Research, 1996); Kaufmann and Ochumba, op. cit. note 2.

60. Baillie and Groombridge, op. cit. note 2.

61. Ibid; Amanda C.J. Vincent, *The International Trade in Seahorses* (Cambridge, U.K.: TRAFFIC International, 1996).

62. Debra A. Rose, *An Overview of World Trade in Sharks and Other Cartilaginous Fishes* (Cambridge, U.K.: TRAFFIC International, 1996).

63. Holly Reed, "Caviar Trade Threatens Caspian Sea Sturgeon," *TRAFFIC USA Bulletin*, December 1996; Vadim Birstein, "Sturgeons and Paddlefishes: Threatened Fishes in Need of Conservation," *Conservation Biology*, December 1993; Alexander Amstislavskii, "Sturgeon and Salmon on the Verge of Extinction," *Environmental Policy Review*, vol. 5, no. 1, 1991; Baillie and Groombridge, op. cit. note 2.

64. Ibid.

65. One quarter figure calculated by Worldwatch from Baillie and Groombridge, op. cit. note 2; for details on how the figure was arrived at, see note 4.

66. Baillie and Groombridge, op. cit. note 2.

67. Figure for Brazilian birds calculated from Stattersfield et al., op. cit. note 17; primates calculated from Baillie and Groombridge, op. cit. note 2, and from Rylands et al., op. cit. note 30; A.P. Dobson, J.P. Rodriguez, W.M. Roberts, and D.S. Wilcove, "Geographic Distribution of Endangered Species in the United States," *Science*, 24 January 1997.

68. Lee Hannah et al., "A Preliminary Inventory of Human Disturbance of World Ecosystems," *Ambio*, July 1994; Lee Hannah, John L. Carr, and Ali Lankerani, "Human Disturbance and Natural Habitat: A Biome Level Analysis of a Global Data Set," *Biodiversity and Conservation*, vol. 4, 1995; Nels Johnson, *Biodiversity in the Balance: Approaches to Setting Geographic Conservation Priorities* (Washington, DC: Biodiversity Support Program, 1995).

69. Hannah et al., op. cit. note 68; Hannah, Carr, and Lankerani, op. cit. note 68.

70. Hannah et al., op. cit. note 68; Dirk Bryant, Daniel Nielsen, and Laura Tangley, *The Last Frontier Forests: Ecosystems and Economies on the Edge* (Washington, DC: World Resources Institute (WRI), 1997).

71. Peter Brazaitis, Myrna E. Watanabe, and George Amato, "The Caiman Trade," *Scientific American*, March 1998; Appalachian large mammals from E. Raymond Hall, *The Mammals of North America* (New York: John Wiley & Sons, 1981).

72. Caribbean mammal extinctions from McPhee and Flemming, op. cit. note 40.

73. Lack of chemical contamination for current threatened species from Baillie and Groombridge, op. cit. note 2.

74. Jennifer Mitchell, "Nowhere to Hide: The Global Spread of High-Risk Synthetic Chemicals," *World Watch*, March/April 1997; Klaffke, op. cit. note 1.

75. Chris Bright, "Tracking the Ecology of Climate Change," in Lester R. Brown et al., op. cit. note 7.

76. Decline of U.S. mussels from James D. Williams and Richard J. Neves, "Freshwater Mussels: A Neglected and Declining Aquatic Resource," in LaRoe et al., op. cit note 10; information from Mobile River snails from Arthur E. Bogan, J. Malcolm Pierson, and Paul Hartfield, "Decline in the Freshwater Gastropod Fauna in the Mobile Bay Basin," in LaRoe et al., op. cit. note 10; information on wolfsnail and flatworm from Bruce Stein and Stephanie Flack, eds., *America's Least Wanted: Alien Species Invasions of U.S. Ecosystems* (Arlington, VA: Nature Conservancy, 1996), and from Baillie and Groombridge, op. cit. note 2.

77. Defenders of Wildlife, "How the Endangered Species Act Works," <http://www.defenders.org/esahow.html>, and "ESA Success Stories," <http://www.defenders.org/esasucc.html>, viewed 8 April 1998.

78. The situation in Darién province, Panama, from personal observations of author; population figures from *Panamá en Cifras* (Panama City: Contraloria General de la República, 1971 and 1996 editions); forest cover figures from Caroline S. Harcourt and Jeffrey Sayer, eds., *The Conservation Atlas of Tropical Forests: The Americas* (New York: Simon & Schuster, 1996); social factors behind deforestation from Charlotte Elton, ed., *Panamá: Evaluación de la Sostenibilidad Nacional* (Panama City: Centro de Estudios y Acción Social Panameño, 1997) and from Stanley Heckadon Moreno, *Cuando Se Acaban Los Montes* (Panama City: Editorial Universitaria Panamá, 1983).

79. Global protected area total from Catherine Baden-Daintree, "Protected Area Boom," *Oryx*, April 1997.

80. Total for conservation-dependent African megafauna calculated from Baillie and Groombridge, op. cit. note 2.

81. Increase in protected areas coverage from 1990 to 1995 from Baden-Daintree, op. cit. note 79; the lack of protection given to tropical wetlands and other freshwater habitats is commented on by Norman Myers, "The Rich Diversity of Biodiversity Issues," in Reaka-Kudla, Wilson, and Wilson, op. cit. note 6.

82. Johnson, op. cit. note 68; Stattersfield et al., op. cit. note 17; Eric Dinerstein et al., *Una Evaluación del Estado de Conservación de las Ecoregiones Terrestres de América Latina y El Caribe*, (Washington, DC: World Bank, 1995); Nature Conservancy, *Designing A Geography of Hope: Guidelines for Ecoregion-Based Conservation in Nature Conservancy* (Arlington, VA: Nature Conservancy, 1997).

83. Carlos F. Guidon, "The Importance of Forest Fragments to the Maintenance of Regional Biodiversity in Costa Rica," in John Schelhas and Russell Greenberg, eds., *Forest Patches in Tropical Landscapes* (Washington, DC: Island Press, 1996).

84. World Conservation Monitoring Centre, "1996 Global Protected Areas Summary Statistics," <http://www.wcmc.org.uk/protected_areas/data/summstat.html>, viewed 15 April 1998; Lee Hannah, *African People, African Parks: An Evaluation of Development Initiatives as a Means of Improving Protected Area Conservation in Africa* (Washington, DC: Conservation International, 1992); Carel van Schaik, John Terborgh, and Barbara L. Dugelby, "The Silent Crisis: The State of Rain Forest Nature Preserves," in Randall Kramer, Carel van Schaik, and Julie Johnson, eds., *Last Stand: Protected Areas and the Defense of Tropical Biodiversity* (New York: Oxford University Press, 1997).

85. De-gazetting of Indian nature reserves from Sanjay Kumar, "Mining Digs Deep Into India's Wildlife Refuges," *New Scientist*, 26 August 1995.

86. Daniel H. Janzen, "Wildland Biodiversity Management in the Tropics," in Reaka-Kudla, Wilson, and Wilson, op. cit. note 6.

87. Janice Rosenberg and Gary P. Nabhan, "Where Ancient Stories Guide Children Home," *Natural History*, October 1997; Gary Paul Nabhan, *Cultures of Habitat* (Washington, DC: Counterpoint Press, 1997); World Wildlife Fund (WWF), *The Biodiversity Collection: A Review of Biodiversity Resources for Educators* (Washington, DC: 1998); Indonesian theater from Nancy L. Peluso, *Rich Forests, Poor People* (Berkeley, CA: University of California Press, 1993); Uganda theater from Stephen Rwangyezi and Paul Woomer, "Promoting Environmental Awareness Through Performance Education," *Nature and Resources*, vol. 31, no. 4, 1995.

88. The extent of overlap between protected areas and traditionally managed lands is detailed by Marcus Colchester, *Salvaging Nature: Indigenous Peoples, Protected Areas and Biodiversity Conservation* (Geneva: United Nations Research Institute for Social Development (UNRISD), 1994); Aiah

R. Lebbie and Raymond P. Guries, "Ethnobotanical Value and Conservation of Sacred Groves of the Kpaa Mende in Sierra Leone," *Economic Botany*, vol. 49, no. 3 1995; Mac Chapin, "The Co-existence of Indigenous Peoples and Environments in Central America," *National Geographic Research and Exploration*, vol. 8, no. 2, 1992 (map supplement); Bernard Nietschmann, "Protecting Indigenous Coral Reefs and Sea Territories, Miskito Coast, RAAN, Nicaragua," in Stan Stevens, ed., *Conservation Through Cultural Survival: Indigenous Peoples and Protected Areas* (Washington, DC: Island Press, 1997).

89.   Colchester, op. cit. note 88.

90.   Dale M. Lewis, "The Zambian Way to Africanize Conservation," in Dale Lewis and Nick Carter, eds., *Voices from Africa: Local Perspectives on Conservation* (Washington, DC: WWF, 1993).

91.   Lewis, op. cit. note 90; Cheri Sugal, "The Price of Habitat," *World Watch*, May/June 1997.

92.   Quote from Jeffrey Sayer, *Rainforest Buffer Zones: Guidelines for Protected Area Managers* (Gland, Switzerland: IUCN, 1991); percentage of inhabited protected areas from John A. Dixon and Paul B. Sherman, *Economics of Protected Areas: A New Look at Benefits and Costs* (London: Earthscan, 1991); integrated protected areas reviewed by Michel P. Pimbert and Jules N. Pretty, *Parks, People and Professionals: Putting "Participation" Into Protected Area Management* (Geneva: UNRISD, 1995).

93.   Stan Stevens, "Annapurna Conservation Area: Empowerment, Conservation, and Development in Nepal," in Stevens, op. cit. note 88; Stanley F. Stevens and Mingma Norbu Sherpa, "Indigenous Peoples and Protected Areas: New Approaches to Conservation in Highland Nepal," in Lawrence S. Hamilton, Daniel P. Bauer, and Helen F. Takeuchi, eds., *Parks, Peaks, and People* (Honolulu, HI: East-West Center, 1993); Michael P. Wells, "A Profile and Interim Assessment of the Annapurna Conservation Area Project, Nepal," in David Western and R. Michael Wright, eds., *Natural Connections: Perspectives in Community-Based Conservation* (Washington, DC: Island Press, 1994); Hubert Job and Manfred Thomaser, "Conservation for Development: Nepal on the Road to an Integrated Policy of Nature Conservation and Development," in *Applied Geography and Development*, Volume 49, (Tübingen, Germany: Institute for Scientific Co-operation, 1997).

94.   Stan Stevens, "Annapurna Conservation Area: Empowerment, Conservation, and Development in Nepal," in Stevens, op. cit. note 88; Stanley F. Stevens and Mingma Norbu Sherpa, "Indigenous Peoples and Protected Areas: New Approaches to Conservation in Highland Nepal," in Lawrence S. Hamilton, Daniel P. Bauer, and Helen F. Takeuchi, eds., *Parks, Peaks, and People* (Honolulu, HI: East-West Center, 1993); Michael P. Wells, "A Profile and Interim Assessment of the Annapurna Conservation Area

Project, Nepal," in David Western and R. Michael Wright, eds., *Natural Connections: Perspectives in Community-Based Conservation* (Washington, DC: Island Press, 1994); Hubert Job and Manfred Thomaser, "Conservation for Development: Nepal on the Road to an Integrated Policy of Nature Conservation and Development," in *Applied Geography and Development*, Volume 49, (Tübingen, Germany: Institute for Scientific Co-operation, 1997).

95. Stan Stevens, "Annapurna Conservation Area: Empowerment, Conservation, and Development in Nepal," in Stevens, op. cit. note 88; Stanley F. Stevens and Mingma Norbu Sherpa, "Indigenous Peoples and Protected Areas: New Approaches to Conservation in Highland Nepal," in Lawrence S. Hamilton, Daniel P. Bauer, and Helen F. Takeuchi, eds., *Parks, Peaks, and People* (Honolulu, HI: East-West Center, 1993); Michael P. Wells, "A Profile and Interim Assessment of the Annapurna Conservation Area Project, Nepal," in David Western and R. Michael Wright, eds., *Natural Connections: Perspectives in Community-Based Conservation* (Washington, DC: Island Press, 1994); Hubert Job and Manfred Thomaser, "Conservation for Development: Nepal on the Road to an Integrated Policy of Nature Conservation and Development," in *Applied Geography and Development*, Volume 49, (Tübingen, Germany: Institute for Scientific Co-operation, 1997).

96. Stan Stevens, "Annapurna Conservation Area: Empowerment, Conservation, and Development in Nepal," in Stevens, op. cit. note 88; Stanley F. Stevens and Mingma Norbu Sherpa, "Indigenous Peoples and Protected Areas: New Approaches to Conservation in Highland Nepal," in Lawrence S. Hamilton, Daniel P. Bauer, and Helen F. Takeuchi, eds., *Parks, Peaks, and People* (Honolulu, HI: East-West Center, 1993); Michael P. Wells, "A Profile and Interim Assessment of the Annapurna Conservation Area Project, Nepal," in David Western and R. Michael Wright, eds., *Natural Connections: Perspectives in Community-Based Conservation* (Washington, DC: Island Press, 1994); Hubert Job and Manfred Thomaser, "Conservation for Development: Nepal on the Road to an Integrated Policy of Nature Conservation and Development," in *Applied Geography and Development*, Volume 49, (Tübingen, Germany: Institute for Scientific Co-operation, 1997).

97. Stan Stevens, "Annapurna Conservation Area: Empowerment, Conservation, and Development in Nepal," in Stevens, op. cit. note 88; Stanley F. Stevens and Mingma Norbu Sherpa, "Indigenous Peoples and Protected Areas: New Approaches to Conservation in Highland Nepal," in Lawrence S. Hamilton, Daniel P. Bauer, and Helen F. Takeuchi, eds., *Parks, Peaks, and People* (Honolulu, HI: East-West Center, 1993); Michael P. Wells, "A Profile and Interim Assessment of the Annapurna Conservation Area Project, Nepal," in David Western and R. Michael Wright, eds., *Natural Connections: Perspectives in Community-Based Conservation* (Washington, DC: Island Press, 1994); Hubert Job and Manfred Thomaser, "Conservation for Development: Nepal on the Road to an Integrated Policy of Nature

Conservation and Development," in *Applied Geography and Development*, Volume 49, (Tübingen, Germany: Institute for Scientific Co-operation, 1997).

98. Katrina Brandon, "Policy and Practical Considerations in Land-Use Strategies for Biodiversity Conservation," in Kramer et al., op. cit. note 84; Andrew J. Noss, "Challenges to Nature Conservation With Community Development in Central African Forests," *Oryx*, October 1997.

99. Population figures from U.N. Population Division, *World Population Projections to 2150* (New York: United Nations Secretariat, 1998); future population projection is using low and high fertility scenarios; Lewis and Carter, op. cit. note 90.

100. Brandon, op. cit. note 98.

101. William J. Snape III, "International Protection: Beyond Human Boundaries," in William J. Snape III, ed., *Biodiversity and the Law* (Washington, DC: Island Press/Defenders of Wildlife, 1996); African elephant information from Robin Sharp, "The African Elephant: Conservation and CITES," *Oryx*, April 1997.

102. Sharp, op. cit. note 101; Susan L. Crowley, "Saving Africa's Elephants: No Easy Answers," *African Wildlife News*, May-June 1997; World Wildlife Fund, "Summary Report on the 1997 CITES Conference," *TRAFFIC USA Bulletin*, August 1997.

103. WWF, "Taiwan Certification Officially Lifted," *TRAFFIC USA Bulletin*, December 1996.

104. Leesteffy Jenkins, "Using Trade Measures to Protect Biodiversity," in Snape, op. cit. note 101; Adam Entous, "WTO Rules Against U.S. on Sea Turtle Protection Law," *Reuters*, 6 April 1998.

105. "International Ban Imposed on Thai Wildlife Trade," *The Nation* (Bangkok), 16 April 1991; "Government Will Not Oppose U.S. Ban on Thai Wildlife Imports," *The Nation* (Bangkok), 5 July 1991.

106. Snape, op. cit. note 101; brown tree snake information from Mark Jaffe, *And No Birds Sing* (New York: Simon & Schuster, 1994).

107. Convention on Biological Diversity Secretariat, <http://www.biodiv. org>, viewed 22 April 1998; IUCN-World Conservation Union, *The Global Environment Facility-From Rio to New Delhi: A Guide for NGOs* (Gland, Switzerland and Cambridge, U.K.: IUCN, 1997); Anatole F. Krattiger et al., eds., *Widening Perspectives on Biodiversity* (Gland, Switzerland: IUCN-World Conservation Union and International Academy of the Environment, 1994).

108. CBD country report numbers from "National Reports," 15 March 1998, <http://www.biodiv.org/nat-repo.html>; IUCN-World Conservation Union, op. cit. note 107; amounts pledged to GEF from Hilary F. French, *Partnership for the Planet: An Environmental Agenda for the United Nations,* Worldwatch Paper 126 (Washington, DC: Worldwatch Institute, July 1995); donor country aid shortfalls from BirdLife International, *New and Additional? Financial Resources for Biodiversity Conservation in Developing Countries 1987–1994* (Bedfordshire, U.K.: Royal Society for the Protection of Birds, 1996).

109. IUCN-World Conservation Union, op. cit. note 107; Scott Hajost and Curtis Fish, "Biodiversity Conservation and International Instruments," in Snape, op. cit. note 101.

110. Neitschmann, op. cit. note 88; Janet N. Abramovitz, "Sustaining the World's Forests," in Lester R. Brown et al., op. cit. note 58; Susan C. Stonich, "Reclaiming the Commons: Grassroots Resistance and Retaliation in Honduras," *Cultural Survival Quarterly,* spring 1996; Hilary F. French, *Investing In the Future: Harnessing Private Capital Flows for Environmentally Sustainable Development,* Worldwatch Paper 139 (Washington, DC: Worldwatch Institute, February 1998).

111. French, op. cit. note 110; BirdLife International, op. cit. note 108.

112. French, op. cit. note 110; BirdLife International, op. cit. note 108.

113. Ibid.

114. Hinrichsen, op. cit. note 57; French op. cit. note 110; Pimbert and Pretty, op. cit. note 92.

115. Douglas W. Yu, Thomas Hendrickson, and Ada Castillo, "Ecotourism and Conservation in Amazonian Perú: Short-term and Long-term Challenges," *Environmental Conservation,* June 1997; Terry de Lacy and Bruce Lawson, "The Uluru/Kakadu Model: Joint Management of Aboriginal-Owned National Parks in Australia," in Stevens, op. cit. note 88.

116. Brazil policy change from Ronald A. Foresta, *Amazonian Conservation in the Age of Development* (Gainesville, FL: University of Florida Press, 1991); examples of other needed reforms from Colchester and Nietschmann, both op. cit. note 88; and from Peter Herlihy, "Indigenous Peoples and Biosphere Reserve Conservation in the Mosquitia Rainforest Corridor, Honduras," in Stevens, ed., op. cit. note 88; conservation funding from BirdLife International, op. cit. note 108; David Malin Roodman, *Paying the Piper: Subsidies, Politics and the Environment,* Worldwatch Paper 133 (Washington, DC: Worldwatch Institute, December 1996); Hajost and Fish, op. cit. note 109.

117. The Indonesian government claims that the total area (primary forest, secondary forest, and scrubland) engulfed by fires was 300,000 hectares, while non-governmental organizations estimate between 750,000 and 2 million hectares were affected. Todd Crowell and Peter Morgan, "The Year the Sky Turned Yellow," *Asiaweek*, 26 December 1997; Nick Brown, "Out of Control: Fires and Forestry in Indonesia," *Trends in Ecology and Evolution*, January 1998; Bruno Manser-Fonds, "Why Are the Forests Burning?" *The Ecologist*, January/February 1998; Nigel Dudley, *The Year the World Caught Fire* (Godalming, U.K.: World Wide Fund for Nature, 1997); U.S. Department of Agriculture (USDA), *Production, Supply and Distribution*, electronic database, Washington, DC, updated March 1998; Center for International Forestry Research, "Indonesian Forests and the Financial Crisis," <http://www.cgiar.org/cifor>, viewed 15 March 1998.

118. USDA, op. cit.; Carol Pierce Colfer, "El Niño's Human Face," *CIFOR News*, September 1997.

119. Abramovitz, op. cit. note 110.

120. Ivette Perfecto, Robert Rice, Russell Greenberg, and Martha Van der Voort, "Shade Coffee: A Disappearing Refuge for Biodiversity," *Bioscience*, September 1996; Rainforest Alliance, "Eco-OK Roasters," *The Canopy*, September/October 1997; Eco-OK Coffee Program news release, <http://www.rainforest-alliance.org/pr_okun.html>, viewed 10 March 1998.

121. Niles Eldredge, *The Miner's Canary: Unraveling the Mysteries of Extinction* (New York: Prentice Hall Press, 1991).

Worldwatch Papers

No. of Copies

_____**Total copies (transfer number to order form on next page)**

PUBLICATION ORDER FORM

_____ *State of the World:* $13.95
The annual book used by journalists, activists, scholars, and policymakers
worldwide to get a clear picture of the environmental problems we face.

_____ **Worldwatch Library: $30.00 (international subscribers $45)**
Receive *State of the World* and all six Worldwatch Papers as they are released
during the calendar year.

_____ *Vital Signs:* **$12.00**
The book of trends that are shaping our future in easy-to-read graph and table
format, with a brief commentary on each trend.

_____ **WORLD WATCH magazine subscription: $20.00 (international airmail $35.00)**
Stay abreast of global environmental trends and issues with our award-winning,
eminently readable bimonthly magazine.

_____ **Worldwatch Database Disk Subscription: $89.00**
Contains global agricultural, energy, economic, environmental, social, and
military indicators from all current Worldwatch publications. Includes a mid-year
update, and free copies of *Vital Signs* and *State of the World* as they are
published. Can be used with Lotus 1-2-3, Quattro Pro, Excel, SuperCalc and
many other spreadsheets.
Check one: _____ **IBM-compatible** _____ **Macintosh**

_____ **Worldwatch Papers—See list on previous page**
Single copy: $5.00 • 2–5: $4.00 ea. • 6–20: $3.00 ea. • 21 or more:
$2.00 ea. (Call Vice President for Communications at (202) 452-1999
for discounts on larger orders.)

$4.00* Shipping and Handling *($8.00 outside North America)*

minimum charge for S&H; call (800) 555-2028 for bulk order S&H

_____ **TOTAL** (U.S. dollars only)

Make check payable to Worldwatch Institute

1776 Massachusetts Ave., NW, Washington, DC 20036-1904 USA

Enclosed is my check or purchase order for U.S. $_____

☐ AMEX ☐ VISA ☐ MasterCard _____
                                   Card Number   Expiration Date

signature

name   **daytime phone #**

address

city   **state**   **zip/country**

phone: (202) 452-1999   fax: (202) 296-7365   e-mail: wwpub@worldwatch.org
website: www.worldwatch.org

Wish to make a tax-deductible contribution? Contact Worldwatch to find out how
your donation can help advance our work.